BANK 2.0

"The impact of the Internet and mobile devices has made the rules in managing channels and how we reach customers a moving target. This book does something that no one I know has been able to do thus far —teach us to re-design our instincts first and then our knowledge about how this moving target will evolve. With the correct instinct, we will be able to respond correctly to the rules as they change. I am very grateful to Brett for putting down to paper the instincts that he has been able to hone over the years. Brett is a true international; he is probably one of the few I know who can draw from personal examples from across Asia, where as much and maybe more innovations are taking place in financial services, as anywhere else in the world."

Emmanuel Daniel,
Chairman, *The Asian Banker Journal*

"Creating more value for customers is a hallmark of successful and growing organisations. But the field of competitive battle has changed. What customers value today is different from what they appreciated years ago, and will be very different once again in the rapidly unfolding future. **BANK** 2.0 brings together Brett King's incomparable view of technology, strategy, customer value and delivering superior service. His insights are a "must read" for anyone who wants to attract and keep customers in the incredible years ahead."

Ron Kaufman,
author and founder of UP Your Service! College

BANK 2.0

HOW CUSTOMER BEHAVIOUR
AND TECHNOLOGY WILL CHANGE
THE FUTURE OF FINANCIAL
SERVICES_

BANK 2.0

HOW CUSTOMER BEHAVIOUR
AND TECHNOLOGY WILL CHANGE
THE FUTURE OF FINANCIAL
SERVICES_

BRETT KING

Marshall Cavendish
Business

Reprinted 2010 (four times), 2011 (three times), 2012

Project editor: Lee Mei Lin
Designer: Steven Tan / Cover art by OpalWorks Co. Ltd

All content information in this book is correct at press time. Care has been taken to trace the ownership of any copyright material contained in the book. Photographs are used with permission and credit given to the photographer or copyright holder.

This edition published 2010 by
Marshall Cavendish Business
An imprint of Marshall Cavendish International
1 New Industrial Road, Singapore 536196

Other Marshall Cavendish Offices
Marshall Cavendish Ltd. PO Box 65829, London EC1P 1NY, UK • Marshall Cavendish Corporation. 99 White Plains Road, Tarrytown NY 10591-9001, USA • Marshall Cavendish International (Thailand) Co Ltd. 253 Asoke, 12th Flr, Sukhumvit 21 Road, Klongtoey Nua, Wattana, Bangkok 10110, Thailand • Marshall Cavendish (Malaysia) Sdn Bhd, Times Subang, Lot 46, Subang Hi-Tech Industrial Park, Batu Tiga, 40000 Shah Alam, Selangor Darul Ehsan, Malaysia

Marshall Cavendish is a trademark of Times Publishing Limited

National Library Board Singapore Cataloguing in Publication Data
King, Brett, 1968-
Bank 2.0 : how customer behaviour and technology will change the future of financial services / Brett King. – Singapore : Marshall Cavendish Business, 2010.

p. cm.
ISBN-13 : 978-981-4302-07-4

1. Banks and banking – Customer services. 2. Bank management. 3. Financial services industry. I. Title.

HG1616.C87
332.17 – dc22   OCN520278739

Printed and bound in Great Britain
by TJ International Ltd, Padstow, Cornwall

To Beck, Hannah, Matthew and Thomas
who are the reason for everything

Contents

PART THREE: The Road Ahead—Beyond Channel

Chapter 9: Deep Impact—Technology and Disruptive Innovation

Chapter 10: Gridless Customer Experience—More Complexity, More Choice

Chapter 11: The Emergence of the Prosumer—Collective Intelligence, Social Networking and Web 2.0

Chapter 12: Future Payments and Cash—RFID, Biometrics, P2P Micropayments, Digital Cash

BANK 2.0

HOW CUSTOMER BEHAVIOUR
AND TECHNOLOGY WILL CHANGE
THE FUTURE OF FINANCIAL
SERVICES_

Preface

A STAGGERING 90 per cent of daily transactions are executed electronically today. Institutions that hold on to the belief that physical branches remain at the core of what the brand does, will not adapt easily to the customer of tomorrow who rarely visits a branch or the customer who sees no need for an over-the-counter transaction with cash or cheques. Those who still classify the Internet, ATM and iPhone applications as "alternative" channels will be playing catch up for the next decade, while intermediaries will increasingly capture niche service opportunities. This is where **BANK** 2.0 starts.

Let us be clear. This book is not for traditional bankers who want to stick to the status quo. This book is designed for managers of change who need to argue change vigorously within the institution and show the hard facts that make such change inevitable. But there is an upside. That is, **BANK** 2.0 will show bankers that such changes, although inevitable, will bring reduction in costs, longer-term and more profitable customer relationships, and will improve the effectiveness of the organisation structure. It just may be extremely painful for those who don't get the future, despite all the benefits.

Deregulation, increasing volatility in capital markets and currency, lower barriers to entry, lower product margins, and headlines around issues such as massive subprime "adjustments" and record "bailout-bonuses" make banks much less of a "blue chip" investment today than they were in days gone by. Customers are also increasingly vocal and brutal when it comes to poor customer affinity within the financial services sector. The recent MoveYourMoney.info movement in the United States was a case in point. Social media tools such as Twitter and blogs are simply increasing the negative buzz around outmoded banking practices.

So if you are in a business environment that hasn't significantly changed its operational model in the last 100 years, where customers are increasingly vocal in their criticisms of your practices and policies, and the margins, profits, management practices and valuations of your organisation are under attack... you might be forgiven for thinking that something needs to change.

Change it will. In the next five years, cheques will be phased out in most developed economies, mobile banking via app phones will represent the fastest growing banking channel in the next two to three years, and branch visits will decline further as time-poor customers struggle with the demands on their schedule. Banks will be challenged by customers who simply no longer see the need for KYC (Know-Your-Customer) compliance based on physical paperwork, or who can't spare the time to visit a branch. Within 10 years, most of the traditional media we use today, namely TV, newspapers, magazines and so forth will have changed forever. Direct mail as a marketing medium has already failed, TV commercials, physical newspapers and most other forms of advertising will have disappeared by 2020.

We have learned that channels are just about ways to reach and engage customers, and despite our best efforts, customers wanted to use these channels *their* way—we had very limited success in "training" them to use the channels our way. In terms of banking, customers still needed to get from point A to point B, but they are looking for smarter, simpler ways of doing that.

The research for this book is 10 years in the making. In the end we had audited, measured and experimented with almost every possible combination of channels, products and tools that customers and bankers alike have access to.

What we have learned and what we have selected as case studies to put into this book is not really about technology or about any one channel. It is about customers and what they need from their institution.

What we've seen, without banks even realising it is happening, is that a customer revolution of sorts is taking place. Organisation structures within the majority of banks still generally reinforce a branch-led approach to banking, while channels such as call centres, the Internet and ATM are often relegated to "'alternative channels", serviced by non-strategic silos. There is little thought given to integrating such channels as the mini technology empires, product teams and marketers reinforce silos due to antiquated KPIs (Key Performance Indicators). New technologies such as mobile-based Internet, social networking and cashless payment solutions are also emerging to create further havoc with the traditional models.

Thus, in the year 2010, we've reached an epiphany—the branch, the Internet, the call centre, IVR systems, ATM, kiosks, Web 2.0, in fact, every channel and mechanism we've got at our disposal is simply just part of the retail mix.

Customers choose those channels and interactions that get them to their desired solution in the quickest, most efficient manner. They don't choose the branch because they get "better service" *unless* they need a human face as an element of that specific solution, and then, increasingly, under protest because we are so time-poor these days. They don't choose Internet banking because the bank has some great functionality, tools or technology in place. They certainly don't choose the call centre because they will get the best answers to their questions. They choose a channel most of the time because they just don't have a choice. Customers simply want a great banking service, tailored to their needs and in a timeframe that works. In other words, a total customer experience.

BANK 2.0 allows us to show you how.

Acknowledgements

I WOULD like to pass on my thanks to the following people who assisted me in pulling together this book.

Firstly to Michael Armstrong who contributed frequently, and put in a big effort on the chapter on ATM and reviewed many of the chapters with great feedback and insights. Michael started as a client, and is now a valued friend and colleague. I couldn't have done this without him.

To the team at HSBC, including Louisa Cheang, Peter Brooks, Martin Rawling, Christina Yung and Matthew Dooley. Also, Steve Townend at MoBank, David Brierely and the team at QlikTech, Ron Kaufman at UP Your Service! College, Ben May at OnlineFactor, Dave and the team at Heath Wallace, and the various bloggers, researchers and others who allowed me to use their insights in this collective work. To David Cavell who gave great insight into the challenges of branch network evolution.

To Chris Skinner, Mike Walsh, Richard Nearn and Richard Petty for their moments of clarity and advice in respect of the publication strategy. To Sean Clifford who is virtually an investor in this book with his "logistical" support over many months; Sean, you are a great friend, and a better advisor.

My PR powerhouse Kirstin Myers from Globond who carried me along with her energy and enthusiasm on more than one occasion.

To the team at Marshall Cavendish, but particularly Lee Mei Lin, who worked hard on getting this book to print in a timely fashion and kept me focused on the end goal. Thanks to Saw Puay Lim, my editor, who kept me from getting bogged down in detail.

But most of all to my wife and children who put up with me neglecting them for weeks on end as the deadline drew nearer.

ntroduction:
Out With the Old, In With the New

by Michael Armstrong, former Senior Manager, Customer Propositions,
HSBC Asia Pacific

WHEN Brett asked me to be involved in this project from a content, advisory and review perspective, I was delighted to contribute. Ten years ago when we started our association, the Internet and such things were still considered breaking ground for most of the bankers I knew. The fact is that when we applied the right thinking we had some astonishing results, and most of my colleagues in the retail banking space still chose to ignore those results as statistical anomalies rather than see them as trends representing changing customer behaviour. While I still believe there are elements of commercial banking that won't change because of regulation and other issues, the fact is that life is simply getting more complicated for bankers.

The first major technological innovation in banking was the ATM over 40 years ago, and up until the early 1990s this was really the only customer-facing technology that existed. Following this came phone banking, IVR systems, the advent of the Internet, increasing complex customer database mining, and remote distribution channels.

What came first in this chicken and egg argument—technology or customer demand for technology? Does technology change customer behaviour or is it customer demand that influences technology in banking? In my early years in banking, the mantra often appeared to be "build it and they will come".

When I was with Citibank in Australia during the early 1990s, we created a futuristic "Video Banking Centre". The VBC booths enabled customers to bank without needing a physical teller as we had tele-operators sitting 20 kilometres away in a call centre ready to assist their every need. It was a very advanced technological solution for banking at that time, but we had built it without actually asking our customers what problem we were really trying to solve. Most importantly, we hadn't asked them if they would actually want to use it. We had let our technicians run the project and assumed once we built it the customers would come.

We had not done our basic homework on customer behaviour. We had not factored in our positioning, physically and within the banking marketplace. The VBC was located in a street that already had four established national banks that were fully staffed. This presented the customer with a choice between being welcomed into a branch by humans, or entering a self-service technology mall.

Does this mean this type of technology would never work? Absolutely not! In another location, with a different competitor environment, with a different client segment, the VBC could have been a big hit—in fact, in some ways it was ahead of its time. The biggest lesson for us, then, and even more relevant today, is that technology use is determined by our customers, not by banks.

The purpose of this book is to provide insight into HOW technology should work for the business in its aim to make its products and services relevant and accessible to its customers, instead of a fixed sunk cost that sits heavily on a bank's P&L statement.

Can You Build Relationships Through Technology?

Traditionally, we have thought that relationships needed to be built on a face-to-face basis with many banks establishing shop fronts or even "megastores" to serve both high-end and mass market clients in order to sell complicated, advisory type of services and to cement the relationship.

Countering this is a belief in the new information economy that relationships can be established and nurtured through electronic means. The younger generation is more likely to find and trust information received through Facebook and Twitter than they are through traditional media. Our research shows that strong relationships can be developed using technology.

The key, as in any relationship, is the establishment of trust. How can we achieve this? Brett devotes a chapter to the emergence of the Prosumer and Web 2.0, aka "social networking" to discover what hype is and what the reality is.

The Customer Experience

As the example I shared with the Citibank VBC shows, unless the customer experience is mapped out, tested and refined, the process will not work and your customer will walk away. This can be as simple as where the LOGIN button is located on your homepage, or as complex as the efficiency of your payment gateway.

Customer experience is becoming the key differentiator in financial services. But like any new positioning of the brand, you have to think about who your customers are. As the VBC story illustrates, a distribution strategy does not equal customer experience, just as a branch is not your only measure of the customer experience in today's increasingly complex world of interactions. The fact is that today many banks might have a great branch experience, but if you drop the ball, so to speak, on another channel, then all that good work is immediately for nought. The concept that the brand can live or die purely on the branch experience has to be abandoned right here and now. Customers simply don't think like this anymore, so neither should the bank.

Making the Case with Senior Management

It is all very well to explore these issues intellectually, but what does it mean for you and how can it help you practically with your work? In writing this book, Brett is attempting to show HOW to present, WHAT to present and WHY. What numbers should you use? What precedents are there already? Is it a strategic or tactical case? What are the risks?

Brett's approach is to keep it informal while being informative. This is a book written by a very experienced practitioner for practitioners and industry insiders, but equally, customers will get some benefit out of understanding the way their banking experience will change.

In terms of how to approach this book, if you are a channel manager, then by all means go straight to Part Two which discusses channel improvement and see if you can build some prototypes or proof of concepts based on the ideas. Do note, however, that the two chapters in Part One are on how customer behaviour is changing the rules and how customer experience is part of the arsenal of responses for the bank now. Part Three of the book focuses on the future. We are not getting into science fiction here; most of what is discussed is possible or probable in the next five to ten years. You need to be ready for these disruptive technologies and need to start thinking about the implications to your organisation today.

Finally, try to keep an open mind. While banking and retail financial services are generally considered a traditional business arena, the fact is that there have been more changes in this sphere in the last 10 to 15 years than in the preceding 100 years. There are two possible approaches to these changes and the changes that are yet to impact the business in the coming decade. The first is to figure out how to adapt to these challenges and try to benefit from them constructively. The second approach is to insist that lots of people talk about change, but you argue strongly that retail banking in reality just doesn't change and it is the same today as it was 40 years ago and customers still want to use banks the same way they always have.

There will be an ongoing role for a few small banks taking the traditional approach, but for the majority of banks the future is a lot more uncertain. If I had told you 15 years ago that Google would be one of the largest global brands, you would have said "who?" If five years ago I had told you the total number of active Facebook users today would be greater than the physical population of about half the countries in the world, you would have said "what?" Thus, if I tell you today that in the next 10 years a new major brand that leads with technology is likely to dominate the retail banking space based on innovation around customer experience, you are probably going to say "who?"

It is very possible that some enterprising corporate incubator will figure out the customer behaviour secret and construct a purpose-built retail bank that takes off like Twitter globally. The fact that a banking licence is needed is hardly a hurdle—there are plenty of cheap banks to buy today, as Richard Branson recently discovered. The Internet, ATM, call centres and app phones have become mainstream for customers, while banks still classify these as "alternative channels" and maintain an organisation structure where Branch dominates thinking.

BANK 2.0 may very well predict the end of banking as we know it.

Part 01

Changes in Customer Behaviour

1 What the Internet and "CrackBerry" Have Taught Customers

RECENTLY I took a six-month sabbatical to work on this book. For the first week of my musings I turned off my BlackBerry, my second mobile, and only periodically checked my email. A few days later, I got an urgent instant message from my personal assistant just wanting to make sure I was alright as the team had not heard from me in a few days. I had trained my customers and staff to know that whenever they needed me, I would be online and transmitting. If you are not the same these days, you are pretty much an exception.

As it turns out, my experience is not unique. Many users of such pervasive technologies are finding it increasingly difficult to detach themselves from such *always on* access and service, either because of demands from their employers or clients for uninterrupted access[1] or worse because of addictions to connectivity.[2] This almost compulsive need to stay connected is just one of the side effects of the information age.[3]

What has produced this shift? Some would argue that this is a by-product of competition, globalisation and a maturing market, and that is definitely part of the story. However, just as

Figure 1.1 Are you addicted to your CrackBerry?
(Credit: Happy Worker Inc.)

critical is the fact that customers *expect* more timely and efficient service from service providers and are indeed demanding it. We are in the instant access and gratification age. Here are just some of the services or products one can access or purchase instantly online, through the phone, or some other electronic (non-person'd) channel that we did not have instant access to just 10 years ago without talking to an agent, intermediary or service representative:

> Airline tickets, hotel reservations, movie tickets, flowers, restaurant bookings, books, songs (CDs, MP3s, Clips, etc), stock trades, travel insurance, golf clubs, shoes, clothes, dating services, electronic goods, fast-food or gourmet meals delivered to your door, education and training, movies (DVDs, downloads, etc), computer software, personal loans, credit reference checks, credit cards, car service and repair booking services, portfolio management and optimisation, CV/resume production and editing, photo printing services, diet management and nutrition services, etc, etc, etc.

The customers of the information or digital age have been empowered by greater choice, greater access, and better, faster, more efficient modes of delivery and service.

There are two major factors at work here in creating a behavioural change, namely the **psychological impact** of the information age and the associated innovations, and the **process of diffusion** (of innovations). Each of these factors contributes to create a paradigm shift in the way financial institutions need to think about service and engagement of customers.

There are **three phases of disruption** that constitute this behavioural paradigm shift with respect to consumers—the arrival of the Internet, the emergence of the smart device and the move towards mobile payments. We'll look at those too.

Psychological Impact

From a psychological point of view a few things are occurring here. To understand the first element, we can revisit one of the foundation pieces of work in respect of the theory of motivation—that of Maslow's Hierarchy of Needs.[4] Maslow studied exemplary people such as Albert Einstein, Jane Addams and Eleanor Roosevelt, and determined the hierarchical progression of the individual, essentially what amounts to a theory of positive motivation and personal development. Maslow's Hierarchy of Needs (Figure 1.2) is often depicted as a pyramid consisting of five levels. The four lower levels are grouped together as being associated with *physiological* needs, while the top level is termed growth needs associated with *psychological* needs. Deficiency needs are met first, then growth needs drive an individual to pursue personal advancement and development.

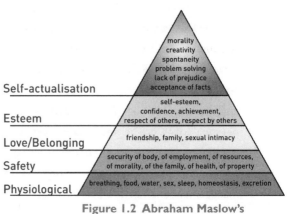

Figure 1.2 Abraham Maslow's Hierachy of Needs, circa 1943

The growth of newer technologies, more efficient service paths and ways to meet our self-actualisation "needs" have shifted the way we value our time, set expectations and perceive ourselves in our environment. For example, we understand through the introduction of new communications channels that if we can do something via phone or online, we are essentially wasting our time by persisting with a traditional interaction that is far less time efficient. This, in turn, increases our self-esteem because we are using our time more wisely. Secondly, the execution of a transaction or a purchase *without* the assistance of a person, as long as it works well, gives us a feeling of control and self-achievement that cannot be achieved in a traditional interaction. Let me illustrate.

Take a mortgage proposition from the 1970s in middle-America. Let's say you wanted to purchase a family home, but needed a mortgage from the bank to accomplish that. In those days, you would need to wander down to the local bank, make an appointment with the manager, and then prepare yourself for an intense grovelling session to see if you would get the bank manager's approval. If the manager liked you, or your business was strategic to the bank, then you might get an offer, but you had zero control of rates, fees and such, as the bank was totally in control. This might have led a customer to feel helpless, even more so if the application was rejected. In this environment, we trained our customers to believe that *they needed the bank more than the bank needed them*, and that they would be rewarded only if they interacted with the bank successfully on the bank's terms. These days, the customer has much greater control over this type of interaction and is not dependent on a limited set of providers, and so he is empowered.

In 2008, the biggest seller of home loans in the United States was Countrywide, recently acquired by Bank of America for US$4.1 billion in stock. Countrywide has more than nine million home mortgages on its books, almost all of which since 2000 have originated online.[5] As a customer, I find organisations such as Countrywide give me access to an almost unlimited choice of financing options for my home. Rates are super competitive and approvals are generally given within 24 hours. I am not required to "front up" in person, or beg for consideration to a potentially biased or partial bank manager. I am in control.

Ostrich-like detractors may jump in here and argue that those who register their interest in mortgage products online, for example, are not generally those most attractive from a demographic point of view, and are probably first home buyers with low incomes, etc. However, anyone making such assertions would be showing their ignorance of the facts. The demographics are crystal clear. Generation Xers (born 1964–75) and baby boomers (born 1946–63) are actually the most likely candidates to research mortgages online.[6] Savings from online mortgage offerings also abound.[7] MyRate, a successful Australian online mortgage provider which is backed by ING, claims it can save borrowers about A$80,000 on an A$300,000, 30-year loan because of the savings the online channel produces. Traditional players have no need to feel under threat by these advances though, as the HSBC Mortgage case study on the following pages shows.

The HSBC Experience
Online Mortgage Lead Generation

With the launch of HSBC's online presence in force in 2000, the bank sought to have a platform for initiating sales opportunities online. The bank contracted Ion Global (previously called Web Connections) to design a world-class public website to parallel their foray into online banking. The site became a benchmark for the group globally, with the red gel-bar or lozenge design appearing in many HSBC country sites around the world.

At the time the consensus was that you could not sell complex products such as mortgages online, and that online was better suited primarily to products you could pay for with a credit card. We were focused purely on the transaction itself as the key element of e-commerce. It wasn't until sometime later that we discovered the benefit of lead generation through Web for offline execution.

It was in this environment that HSBC tentatively introduced a mortgage product page for their retail business in Hong Kong. At first this site received very few visitors, but slowly visits or eyeballs for the mortgage site started to increase as Internet adoption rose. Hong Kong, as a property market, is one of the most sophisticated property markets on the planet, with a significant proportion of its GDP and economic activity generated through investments in property.

Within the first two to three years of the mortgage website's operation, HSBC was successfully generating leads through it. On average by 2002, some eight to ten leads per month were coming through to the DFV (Direct Financial Services) Service Centre via the website.

The user experience, however, was somewhat cumbersome with a five-step process requiring information such as the potential value of the property, the location of the property, your average monthly income and expenditures, how much you wanted to borrow and your personal identification details, before you could even submit a request for more information or ask for an appointment.

In 2002, HSBC embarked on a revamp of the mortgage website, looking to engage customers online better and facilitate more leads for follow-up. The site revamp took six months to complete and involved mortgage product specialists, members of the Direct Channel Development/Internet and Self-Service banking team, IT specialists and the third-party provider Modem Media, which had been contracted to provide design assistance.

The redesign team determined that lead generations could be improved by simplifying the customer experience and not asking them for irrelevant information that was simply not required just to make an appointment. The biggest challenge in this environment was helping the legal and compliance team to understand the difference between the online "make an appointment" form, and an actual application form and the significantly different compliance requirements of the online tool. Once that challenge was met, there needed to be a specific department that could receive those leads and make sure these were followed up quickly and effectively.

The redesigned mortgage products section of the HSBC Hong Kong website was launched in November 2002. No marketing launch accompanied the site, nor was there any online advertising or any traditional media used to promote the newly redesigned site. It was just turned on. The results give us some insight into customer behaviour versus actual channel performance and lead generation.

In the first week, the new site generated **178 leads** through the "make an appointment" contact form (Figure 1.3, p.30), and by the end of the month, almost **800 leads** had been generated. By mid way through 2003, just six months since the site relaunch, more than HK$20 million (US$2.5 million) of mortgage business had been generated from Web leads. In addition, every other measure of site activity available at the time (number of visits/hits, length of stay, etc.) had improved and in most cases doubled during the six months after the launch.

The lead generation example, however, showed us something much more significant in understanding customer behaviour. In the first month of the new site being activated, **traffic was roughly the same** as the previous months, and yet the new site generated close to 800 leads (that's an 8,000 per cent increase!), compared with just 10 leads in the month preceding the launch. As there was no marketing or promotion of the site, it is reasonable to assume that those visitors or "prospects" were already coming to the site to research mortgage products, but they were not using the old lead generation tool because it was too complex. By simplifying the tool, the impact was immediate and overwhelming.

It wasn't a case of build it and they will come, because they were already there. It is simply a case of "think of how the customer behaves", and you will get the benefit.

Figure 1.3 HSBC's Online Mortgage Lead Generation Tool

Make an Appointment

Let us know how to reach you and we'll be in touch to arrange a date and time.

I prefer to be contacted by:

○ Phone   ○ Email

Name:

Email Address:

Contact Number:

Go

So let us look at the psychological influences that these technology and competitive choices give me as an individual. **I am in control** and if the mortgage provider's offer doesn't meet my expectations, I walk away. I have an abundance of choices, and **I am better informed** because of access to extensive informational resources. **I get better deals** because service providers have to work harder to get my business, and **I save money** because the margins have been squeezed through better delivery methods and more competition. **I get a better quality solution** because mortgage products fit my needs much more precisely than the one-size-fits-all solutions that I was restricted to previously. How do I feel about this environment as a consumer, compared with the consumer of the 1970s example? In terms of Maslow's hierarchy, I associate these positive changes as personal development and an improvement in the perception of self. I am more motivated and feel better about myself. **I am happier and in control.**

Over time my overall expectations of my service providers in the finance sector have been lifted to where I now *expect* an element of self-control, efficiency and choice that I didn't have available to me previously. This then moves from being a nice change of pace to becoming a driver of choice and selection, and I penalise providers who aren't able to provide me with this flexibility and level of control and empowerment.

Increased choices in customer touchpoints or channels have given me control about when and how I interact with my bank. Improved information has informed me better so I can get a better deal and push my provider harder. I am empowered and I feel good about that change. This is positive progress.

Process of Diffusion

We'll talk more about this in later chapters, but the other factor in the shifting behaviour of customers is the increasing acceptance of technology and innovation in our daily lives. At the dawn of the 20th century, a bunch of fundamental new technologies were coming, or had recently emerged, onto the scene, namely electricity (1873), the telephone (1876), the automobile (1886), the radio (1906), and in 1903, the aeroplane. This was the dawn of a new age in industrialisation and innovation that caused leaders of the world to claim these improvements would usher in a new age of peace and prosperity. However, these new technologies generally took a significant length of time to reach mass market adoption rates, generally measured in decades.

The physical limitations associated with distribution were a factor in these slow adoption rates, along with the restrictions in knowledge flow and assimilation of technology across geographical boundaries. Another limitation in adoption was purely that such new technologies did not have the advantages of mass production and marketing we have today and were generally time- and resource-expensive to implement. As time passed, adoption rates improved as accessibility to these innovations improved and new production methods emerged. Organisations such as General Electric, Westinghouse, HMV and others also assisted in later decades, through their global presence and faster time-to-market.

By the late 1960s, Moore's Law had kicked into gear, miniaturisation and the "'tronics" fad were leading to an increasing appetite for new gadgets and devices. In the late 60s, TV commercials and print advertisements often touted a science fiction-like future for consumers that was just decades away. Technology and innovation were capturing the imagination of society.

In 1975, IBM invented the personal computer. It wouldn't be launched until a few years later, but it just showed how far technology had come in the three decades since the chairman of IBM had envisaged that there would be a total market globally for only five computers.

Within 10 years, the IBM PC (personal computer) and terms such as DOS, Disk Drive and Dot Matrix (printer) were in the common vernacular. By 1995, when Microsoft launched Windows 95, the desktop computer was a global phenomenon accessible to more than 90 per cent of the world's population and with adoption rates of more than 25 per cent in most of the developed economies of the world. The launch of the cell phone in 1983 by Motorola set the pace at which consumers were being bombarded with new and revolutionary technologies. Then in 1991, the Internet burst onto the world scene. The Web was commercialised by 1994 and then reached the dizzy heights of the dot.com bubble in 1999.

The rate of diffusion is the speed at which a new idea spreads from one consumer to the next. Adoption is similar to diffusion except that it also deals with the psychological processes an individual goes through, rather than an aggregate market process. What had been steadily happening since the late 1800s is that the rates of technology adoption and diffusion into society have both been getting faster. While the telephone took approximately 50 years to reach critical mass,[8] television took just half that (around 23–25 years), cell phones and PCs about 12–14 years (half again), and then the Internet took just seven years (half again). Ultimately, new technologies and initiatives such as the iPod and Facebook are now being adopted by consumers *en masse* in a period measuring months, not years. As we become more used to technology and innovation, it is taking us less time to adopt these technologies into our lives, and this further encourages innovation and thus, increases the impact on business.

> *"I think there is a world market for maybe five computers."* Thomas Watson, IBM Chairman, 1943

Simply put: If we aren't introducing innovations into the customer experience at the same rate with which customers are adopting these new technologies, we are at a considerable disadvantage and we risk losing our customers as more agile intermediaries and third parties capture the benefit of the innovation. Just because we are "the Bank" doesn't cut it anymore.

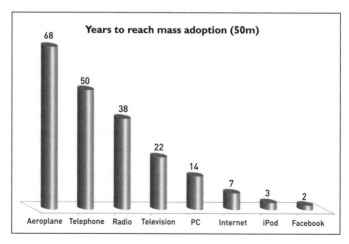

Figure 1.4 Technology adoption rates over the last century

The Three Phases of Behavioural Disruption

There are three stages or phases to the disruption occurring within retail financial services (Figure 1.5). Each stage is disruptive enough to be a "game changer". However, by the time the third phase impacts retail banking around 2015 (or perhaps earlier), the changes will be complete and irreversible.

The first phase occurred with the **arrival of the Internet.** While many banks denied it at the time of the dot.com bubble, the Internet changed forever the way customers accessed their bank and their money. As we discussed in the psychology of customer behaviour, this gave them *control and choice* that was not available previously. Suddenly, customers were thrust into an environment where they could access their money as they wished, when they wished. As Internet banking capability improved, the drive to visit the branch started to diminish and customers began to rely on the new channel as their *primary* access point with the bank for day-to-day transactions. Within just 10 short years, we had gone from 50–60 per cent of transactions done either over the counter at the branch, through ATMs or cash and cheques, to 90 per cent of transactions conducted through the Internet, call centres and ATMs. Game changing…

The second phase is occurring right now. The emergence of the **smart device** or **app phone** such as the iPhone and Google Android enabled phone, is a driver for portable or mobile banking. While many banks may argue about security and the limitations of the screen or the device itself, the fact is we heard exactly the same arguments about Internet banking from those resistant to change within the bank. Already many banks are deploying what amounts to a cashless ATM on a mobile application platform—yes, you can do everything on a mobile phone that you can do on an ATM, *except* withdraw or deposit cash.

Here are a few statistics that support the second phase disruptive model:

- Ninety-three per cent of the US population owns a mobile phone, and 27 per cent of US households are now mobile only.
- Ninety-nine per cent of mobile users view their balances; 90 per cent view transaction details; about $10 billion of funds have

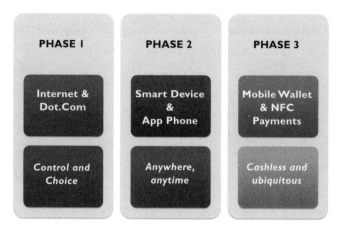

PHASE 1	PHASE 2	PHASE 3
Internet & Dot.Com	Smart Device & App Phone	Mobile Wallet & NFC Payments
Control and Choice	*Anywhere, anytime*	*Cashless and ubiquitous*

Figure 1.5 The three phases of customer-behavioural-led disruption

been moved via mobile transfers/bill pay; 15 million location-based searches are being performed (annual run rate).

- New mobile banking customers at Bank of America (BofA) from July to September 2009: 150,000 (Sep); 210,000 (Aug); 220,000 (July) (Doug Brown, BofA).
- More than 50 per cent of iPhone users have used mobile banking in the past 30 days (Javelin Strategy).
- Thirty-three per cent of mobile banking users monitor accounts daily, 80 per cent weekly (Javelin Strategy).

So, if you didn't need physical cash, what would happen then? This is the third phase—when we move to **mobile payments** on a broad scale. NFC-based (Near Field Communication) mobile wallets and stored value card micropayments are already here, but more is to come. The third phase also involves the convergence of your mobile phone and your credit/debit card, which is a logical technical step in the next five years. When these changes occur, our need for cash will reduce rapidly and the disruption will be far-reaching.

In the UK, 43 per cent of payments are done by debit card and 23 per cent by credit card.[9] Cash still makes up 32 per cent of payments (Figure 1.6 overleaf), but as a percentage of the whole, it continues to reduce. Cheques make up just over 2 per cent of payments these days,

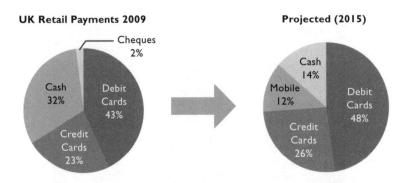

Figure 1.6 Changes expected in UK retail payments, 2006–2015

so it is not hard to see these disappear entirely. As the growth of debit card transactions swells further (not hard to imagine if contactless payment capability is built into your mobile) and other mobile payments like person-to-person (P2P) are enabled on your phone, this will further reduce legacy payment methods. It is not unimaginable to see a split of 85 per cent of UK payments done by mobile/card, and 15 per cent by cash in the next five years. In markets such as Japan, Korea and Hong Kong, the requirement for cash may be even less compared with mobile payments.

There are the great unbanked who don't yet have a bank account and who currently rely heavily on cash, but as we will see with M-PESA and G-Cash in Chapter 6, this is hardly a hurdle for mobile cash and payments. The success of the Octopus card in Hong Kong, T-Money in Korea, Edy and Suica in Japan, and other emerging technologies already prove the concept. What would quickly kill the need for cash in its entirety is a technical standard for mobile money that could be adopted globally by network operators and device manufacturers. Even if only 50 per cent of cash transactions are replaced by electronic stored value cards, debit cards and mobile wallets in the next five to ten years, the current ATM and branch infrastructure that supports cash becomes almost untenable from a cost burden perspective. If you no longer need to go to the ATM to withdraw physical cash or currency, then everything you do on the ATM today can be done on your mobile app phone. If branches no longer need to deal with cash, then a large part of the reason for their existence disappears.

HSBC in the United Kingdom has recently announced its intention to stop supporting cheques because usage has declined and there is no ongoing business case to support them. The Board of the UK Payments Council has agreed to set a target date of 31 October 2018 to close the central cheque clearing system. If cheques decline to the point where banks can no longer afford to support them and regulators no longer require the banks to provide support for them:

> "There will be a critical review in 2016 when the Payments Council will decide whether sufficient change has occurred against agreed published criteria, to press ahead to do away with the cheque in 2018. There are many more efficient ways of making payments than by paper in the 21st century, and the time is ripe for the economy as a whole to reap the benefits of its replacement."[10]
>
> Paul Smee, Chief Executive of the UK Payments Council

Just like Internet and mobile device disruption, these are not the inane ramblings of a technovangelist—this is an inevitable conclusion based on technologies already in place that are on their way to becoming the dominant channel of choice or mechanism of engagement. The behavioural adaption of consumers to the Internet and smart devices already indicate that this will likely take hold within the next three to five years.

Let's just say by the time phase three hits, if the retail banks have not adapted, they will be clinically dead. Banks can either own the transaction and payment platform, integrate the technology, OR protest with their last dying gasp of breath that things are not really going to change. "The Branch is Back", "Cash is King", "Cheques will bounce back"—yes, ok. You just keep telling yourself that and see how that works out for you.

Customer Empowerment

These changes, both in regard to psychology and consumer adoption cycles, have empowered and liberated customers, but represent a real threat to industry. As Evans and Wurster discussed in their book *Blown to Bits* (Wurster, 2000), the threat for traditional intermediaries in particular, is that their business faces potential deconstruction if they cannot encapsulate their place in the value chain in new ways by utilising technology and innovation. This is increasingly why traditional intermediaries such as travel agents and stock brokers are facing an impossible task to maintain margins and restrict churn[11] or loss.

Online stock trading, first embraced by Charles Schwab and the likes of E*Trade, was phenomenally successful in the early days of the commercial World Wide Web, and still is. But there was significant resistance from the likes of traditional players such as Merrill Lynch, which regarded e-trading as a threat to their traditional brokerage model.

The difference in approach between the Charles Schwabs of the world and the Merrills of the world is perhaps the essence in identifying how an organisation copes with challenges presented by innovative technologies in the customer experience.

One reaction is to resist the change because it is uncomfortable and potentially "breaks" our traditional view of the world. The alternative reaction is to realise that this is simply the inevitability of momentum and we need to figure out how to capitalise or benefit. Occasionally, such new technologies turn out to be failures

"The do-it-yourself model of investing, centered on Internet trading, should be regarded as a serious threat to Americans' financial lives."
John 'Launny' Steffens, Vice-Chairman, Merrill Lynch, September 1998[12]

(not fads), like 8-Track, BETA video and WAP. This is due to the fact that more often than not, the new technology is surpassed by something better. The lessons we learn in the first generation of the technology, however, are typically invaluable for future applications.

Today, the Korean Stock Exchange owes 90 per cent[13] of its volume to Internet trades, NASDAQ sees more than 60 per cent of its daily trading volume come from ECN (Electronic Communications Networks) and regional exchanges such as the CME (Chicago Mercantile Exchange) achieve more than 80 per cent of their volume due to electronic trades. Between 2006 and 2007, the New York Mercantile Exchange observed an increase in electronic trading volume of 86 per cent,[14] leading to an overall increase in trading volume of 38 per cent. It would appear by any measure that the online trading experience has been successful.

In Hong Kong, HSBC launched their online trading platform in 2001. Today, more than 80 per cent of all HSBC trades are executed via the Internet. If this facility was to be shut down, there is no way the traditional channels of HSBC could cope with even half of this volume of transactions. Meanwhile, the more than 280 brokerage firms that were present in Hong Kong during the late 90s have dwindled to fewer than 80 players.[15] Indeed, Internet trading was a serious threat, but not to consumers, only to traditional brokerage firms who weren't ready to adapt.

The more advantageous of these transformations have empowered customers in ways that a 1950s bank manager could only have had nightmares about. To illustrate, Table 1.1 below lists three common retail banking products and the average approval times for applications, comparing 1980 and 2008.

Table 1.1 Comparison of product application approval times, 1980 vs 2008

PRODUCT	1980	2008
Credit card	14 days	Instant approval
Personal loan	7–14 days	Pre-approved, or 24 hours
Home mortgage	30 days+	24–48 hours

These product application approval times are indicative of the pressure on financial service providers to adapt to the changing expectations of customers and the need to stay competitive. Barriers to entry are lowering, and new innovations in business models are creating pseudo banking services streamed right to your desktop, supermarket or corner 7-Eleven store. We'll talk more about some of these new models in Chapter 6 on mobile banking and in Chapter 11 on social network-based banking. But for now, suffice it to say that the abundance of new channel delivery methods, most significantly through the Internet and mobile, have created new methods of finding information and engaging financial service organisations. The increased availability of information, in particular, has given consumers more control in the purchase phase and has also enabled service providers to reach more markets.

The other benefit apart from speed of execution, of course, is the ability to execute transactions anywhere, anytime. At the start of the dot.com boom, we used to tout the Internet as anywhere, anytime, but the adoption of BlackBerry and the iPhone have really shown us what *anywhere* can mean. As we move forward, handset technology and mobility will become even more important in keeping us connected. Right now you might do 30–50 per cent of your daily emails on your BlackBerry, and you might even surf the Web on your iPhone 3GS to book movie tickets, airline tickets or other stuff. If I had told you 10 years ago you would be doing these things, you would be arguing it was simply not practical on such a small screen.

This very powerful, portable medium is still extremely constrained by the content we serve up to our customers. We can't expect our customers to access our normal websites designed for PCs and interact via a small mobile touch screen or RIM (Research In Motion) keyboard in the same way. We need to develop new ways of delivering this content in a device-context-sensitive manner, understanding the way our customers interact on a device that is much smaller, more personal and interactive. We'll talk about mobile interaction more in Chapters 9 and 10. But suffice it to say, this will be the new Holy Grail in service delivery if we can get it right.

Putting Service Back Into Financial Services

So what we have effectively taught customers is that they can have control, and they have much more choice. As banks, we still like to think that we can dictate the terms of how and when customers work with us, but increasingly our customers are realising they have a myriad of options. No longer will customers stay with us just because we are the first bank they ever took a deposit account with, or because it appears too hard to change. Those protections will no longer be afforded to a service organisation that doesn't *serve* their customers.

The customer of today expects a total customer experience that works for him. The expectation is that the bank is here to provide me with the services I need. I should not need to justify my worth or suitability as a customer to get those services because if you don't want to work with me I'll walk across the street, or click on the next competitor's website. I want to be in control, and when I need it, I expect rapid and seamless delivery. Don't ask me to fill out an application form with details you already have. I'm not here to work for you; you are here to work for me. Don't dictate to me that I have to go to the branch to do this because I now know that is simply not necessary for a progressive financial services provider with the right systems in place. Understand me, so that you will know what I need before I do. When you recommend a solution to me, don't treat me like a novice; be prepared for me to be well informed and know more about the alternatives than your staff. Tell me why you are recommending this product, and how it fits my needs. **Deliver to my criteria. I'm the customer. It's my total customer experience that matters.**

KEY LESSONS

Customer behaviour is rapidly changing due to two key factors, namely the psychology of self-actualisation, and technology innovation and adoption—otherwise known as diffusion. Banks can either try to reinforce traditional mechanisms and behaviours, or they can anticipate changing behaviour and build accordingly.

The pace and rate of behavioural change is speeding up, not slowing down. Thus institutions get less time to react and anticipate the impact of such changes on their business. The longer institutions wait, the bigger the gap between customer expectations and service capability becomes.

There are three key phases to these behavioural changes and we are already at the second phase. The third phase will occur within the next five years and it will be a game changer.

Keywords Countrywide, Standard Chartered, MyRate, Merrill Lynch, Charles Schwab, KOSDAQ, HKEX, Lead Generation, Psychology, Customer Experience

Endnotes

1   Joseph Pisani, "Workplace BlackBerry Use May Spur Lawsuits", 9 July 2008. Retrieved from CNBC: http://www.cnbc.com/id/25586129

2   Carmel Melouney, "BlackBerry users becoming addicted to gadget". *Sunday Telegraph*, 11 May 2008. Retrieved from http://www.news.com.au/technology/story/0,25642,23676081-5014108,00.html

3   http://en.wiki/information_Age

4   A.H. Maslow, *A Theory of Human Motivation*, Psychological Review 50 (1943): 370–96.

5   Countrywide.com

6   "The next generation of mortgage lead generation" (*Cover Report: Online Lending*) – Matt, Coffin, LowerMyBills.com. Additional sources: Forrester Research Inc, Federal Trade Commission

7   "Online mortgage sites offer net gains", *Australasian Business Intelligence*, 18 September 2006

8   For the purpose of this argument, mass market adoption or critical mass is measured as 25 per cent of the population for developed economies such as the US, the UK, France, Germany and Australia, or 100 million persons globally.

9   *APACS* report on UK payments industry

10   Payments Council Press Release – www.paymentscouncil.org.uk

11   Churn refers to customers moving from a service provider within one specific product category to another based on price, value or some other factor.

12   A widely reported quote from Merrill's John "Launny" Steffens during a presentation on the dangers of e-trading at a PC Expo in San Francisco, September 1998

13   KOSDAQ February 2007

14   Press Release: *PRNewswire*, NEW YORK, 14 December 2007 – NYMEX Holdings, Inc. (NYSE: NMX)

15   Hong Kong Securities and Futures Commission

2 Measuring the Customer Experience

CUSTOMERS are annoying. They don't want to do what they're told, and when we give them straightforward procedures designed by our most intelligent compliance officers and in-house corporate lawyers they complain the processes are long-winded, repetitive and unnecessary. Some customers even tell us they've given us all this information before and have the impertinence to suggest they are doing our work for us! Why can't they just do what they are told?

We've spent millions of dollars on these new fangled computer systems, websites and now even this facetube thingy,[1] and they're still not happy. Why, we've even employed hundreds of computer programmers from Bangalore to churn out all these codes to connect our mainframe systems to Internet banking—and what thanks do we get for it? Really, one would think we would simply be better off if we didn't have to deal with the masses at all!

Seriously folks, in the face of competition, rising infrastructure costs and dwindling product margin, financial service providers have been forced to rethink their strategies on how they reach and serve customers. This has created a number of changes in the way institutions operate and measure success and operational efficiency.

In this chapter, let's examine an innovative approach to measurement based on customer experience. When we talk about customer experience, let's be clear. **This is the total experience the customer has with the bank.** Not a customer satisfaction score, nor the experience he has in the "branch". But every channel, every day for every transaction or service— the total experience.

Over the last few years, we've started to see financial service providers respond to this pressure by taking another look at their channels and the way they measure them. Are customers waiting too long to have their calls answered? Would IVR (Interactive Voice Response) systems lower the load for high frequency enquiries? Are ATMs located in the right places and could customers find them? Who is using the Internet, and what do they want to do online with the bank? Do we need to provide product applications online?

Can we integrate the banking experience better into customers' daily lives? For example, by allowing them to sign up for car financing at the dealer, rather than having to come into the bank and do it as a separate transaction. Could we create mini-branches or sales offices at locations and open at times more appropriate for different customer groups?

Some of this analysis was cost-driven, other initiatives were marketing driven. The problem is that these changes have not been creating better customer experiences holistically. It's been hit and miss. Let's examine why.

Firstly, the **channels are still in silos** that discourage sharing of customer learning, and as a result, some of the most remarkable service opportunities go missing. Secondly, the **organisation structure and traditional business models frustrate change.** The most significant problem, however, is that all these changes are happening in isolation of the customer in most cases. **Customers are rarely involved** in the proposed solutions put forward internally within the institution. Let's discuss these three areas.

Channel Silos

Customers don't use channels or products in isolation of one another. Every day customers will interact with our institution in various ways. They might write a cheque and pass it to a third party, visit an ATM to withdraw cash, go online to check if their salary transfer has come in or pay a utility bill, use their credit card to purchase some goods from a retailer, walk into a branch to collect a personal loan application form, or ring up the call centre to see what their credit card balance is or report a lost card. If they are a sophisticated customer or client, they may also trade some stocks, transfer some cash from their Euro forex account to their US dollar account, put a lump sum in a mutual fund, or sign up for a home insurance policy online.

In the early days of the Internet and call centre, it was not uncommon to find Internet banking and call centre processes lagging 24 hours behind the in-bank systems because the "batch" processes only ran overnight to update the alternate channel databases/logs. So if you made a transaction via an ATM or through the branch, it wouldn't show on your online statement or could not be verified via the call centre until the next morning. There were technical challenges to creating an integrated channel infrastructure from a transactional perspective, largely because we were bolting new channels onto mainframe legacy systems that were simply not designed to work in real time. But these days we have middleware technologies and application servers that allow us to break these technical barriers with relative ease.

The problem remains, however, that the owners of these channels internally within the bank rarely, if ever, talk to one another. In fact, in most instances, the different channel owners see one another as competitors for budget dollars, customer mind-share and share-of-wallet. This also spills over to product teams, who regularly compete against one another for customer attention.

To illustrate the silo problem, I'd like to share an experience I had as a customer of a retail bank in Hong Kong in early 2002. At the time I held a Gold Visa credit card, but I had recently been sent an invitation to upgrade to a Platinum Visa credit card, along with a "pre-approved" application form. I was pleased with this and ready to sign up, but hadn't had the time to fax off the application form (why couldn't I do this online, I thought?). About 10 days later I was in a shop buying a US$5,000 Persian rug, and I got a call from the bank. They were querying the transaction because it was an unusual one-off purchase for me to make. I confirmed the purchase and the transaction was authorised. The CSR (customer service representative) added that the reason they were calling was that some Visa Gold cards had recently been compromised, and to be safe, they could reissue me with a new Gold Visa credit card. I could pick it up from my branch in a few days' time.

I agreed to their suggestion and thought it was proactively a positive move, but I asked them to reissue me with a Platinum Visa card instead, as the bank had sent me the pre-approval offer just a few days earlier. There was silence at the end of the line, followed by the CSR telling me, "I'm very sorry sir, the Platinum Visa credit card department is a separate profit centre within the bank. We are not related." I suggested that maybe the CSR could call the Platinum department and explain the situation and ask them to issue the card, and I would fax the application form to them as soon as I returned to the office. The answer was, "I'm sorry sir, I wouldn't even know who to call. I don't even know if they are in our building…"

From a customer experience point of view, this is a disaster. If I wanted a solution, my ONLY choice at this point was to do all the work on resolving this problem myself. I would have to ring the Platinum department and explain the issue, fax through the form and wait for the new card to be delivered. In the meantime I would have to cancel the Gold card myself, and work out how to transfer the balance from the Gold card to the Platinum card. This would probably mean at least one, but probably a couple of trips to the branch. Why?

The problem with this structure is that the primary measure for these business units or profit centres remains the acquisition of new customers and retention of existing customers. The Gold card team would actually be penalised on a performance measurement basis by recommending I take the new Platinum card. There was no incentive to transfer me over to the new product because their numbers would take a hit. It was only in their interest to do everything possible to retain my account within their product silo, regardless of whether this was best for me or not. The business rewards such profit centres for isolating customers, and categorises activities that holistically provide a better all-round service as inefficient, or worse, irrelevant.

The same thing happens frequently with customer channels. Although mostly those teams do not actively set out to isolate customers, they end up ignoring the rest of channel activity as irrelevant to their part of the world. Call centre teams don't talk to Internet teams, branch teams don't talk to call centre teams. IT, PR and marketing teams frequently battle it out for control of the Web channel. Email marketing and push-mobile services are handled on an ad hoc basis resulting in no one taking control of messages that ultimately reach the customer. Legal and compliance teams frequently hinder channel teams from simplifying application processes through new channels because of a conventional view of the world. It's a mess.

If the institution was to step back from the day-to-day operations and actually look at how a customer interacts with them, they would realise that from a product, process and channel perspective, the customer is totally agnostic.

Customers choose the right channel at the right time for them, depending on a number of factors such as time constraints, always-on availability, complexity and the likelihood of a "deal". What they don't do is think something like, "I think I'd like to go to the branch today to process that travel insurance application." They think, "Hey, I forgot to renew my travel insurance and I'm leaving on Friday. Where am I going to get this done before I go?" If they are comfortable with the Web, they might log on right there and then and apply. Alternately, they might ring the call centre and see if they can sign up over the phone. Or they might call their travel agent or visit their airline's website and see if the airline/hotel package they have has some travel insurance deal linked to it.

So why aren't institutions taking a customer behavioural approach to this, instead of building silos in isolation? The main issue is an organisational structure that is still built on the concept of branch banking, rather than a true multichannel approach.

We're not so much looking for channel migration opportunities, as simply looking at which types of transactions work best on which channel, given a set of circumstances the customer might find himself in. Doing this in an integrated fashion so the customer gets an overall view of the institution is far more important than just blasting individual offers down a new pipeline because the technology allows us to do so.

It is time to get out of the mindset that product teams are competing for mind-share of the customer. There has to be better ways of measuring the success of the total institution in serving the customer, rather than just measuring margin on product, or the number of applications that have come in through the door this month. While revenues, application numbers and transaction activity help compare performance year-on-year and against competitors on a balance sheet basis, aggressive, non-traditional competitors are entering the financial services arena without these pre-conceived notions. As early as 1994, Bill Gates made the statement that "Banking is necessary, but banks are not." We recognise that while banking will not go extinct anytime soon, the pressure is on. Institutions will need to adapt and change and find new ways of working, or give up market share. Yes, many of the traditional "functions" of the bank are now being handled by intermediaries, specialist providers and non-bank institutions. Within the **BANK** 2.0 paradigm, this disruption to the traditional model of banking is set only to accelerate.

Organisation Structure

By examining the behaviour of customers, the glaring realisation is that institutions are essentially assuming that customers only ever use one channel at a time to interact with them. Hence, it is not unusual to find a Web team that believes it can take 30–40 per cent of branch traffic and service it online. Likewise, it is not unusual to hear proponents of branch banking telling us "the branch is back", and that the winning strategy is to be investing in more real estate and variations of branch to retain customers. It's also not unusual for customers to receive dozens of direct mail offers, email marketing offers or SMS promotions from different "revenue centres" within the bank, independent of one another.

In 2008, 90–95 per cent of daily transactions are done electronically[2] and in most cases most of the transaction volume comes through direct channels, namely ATM, call centre and the Internet. By February of 2007, HSBC in Hong Kong reported in the *South China Morning Post*[3] that 90 per cent of their daily transactions were through the phone, the Internet or ATM, leaving the rest to branch. RaboBank, FirstDirect, INGDirect and others have been able to operate successfully without any reliance on branch structures. This is not a criticism of branches because we believe that branches will remain an essential part of the future of banking. However, look at the organisation structure of most banks today and you'll see a

HOT TIP: Guaranteed to reduce your call centre and IVR load by at least 15 per cent this year. For the average bank, this means more than US$1 million in savings.

Channel silos cost banks money because they duplicate functionality and services around customer interaction. Here are a few tips to reduce costs and improve customer satisfaction.

- Don't discourage customers from calling you. Research shows that if you can direct customers to the correct call centre number quickly you reduce traffic and costs.
- On every product or transaction page on your website, list the specific call centre number for that type of product/service. This can direct customers to an IVR menu specifically designed for that query, which will reduce call centre load and ensure CSRs are appropriately equipped to answer specific questions.
- Customers are already coming to your website to find the solution, so why not put a list of the most frequent call types, issues or questions in the same area of the site where they look up the telephone number or even better, on the homepage! You can easily reduce call centre traffic by 10–20 per cent this way. Compile this list of "top" service enquiries by checking call centre data for the most frequent call types over the last six months.
- Keep in mind we would have to provide a solution on the site, not just FAQs. There may have to be some process intelligence, but get this right and those customers already going to your website will not need to ring your call centre. Thus, immediate load reduction.
- Remind customers when they withdraw cash from an ATM that their credit card payment is due.
- Ask customers if they'd like their account balance sent by SMS to their mobile phone. This reduces transaction load on the ATM network for customers who withdraw cash and then re-check their account balance.

complete and total lack of understanding of customer behaviour inherent within the organisation chart. It's really quite appalling that the organisation structure of most banks has not caught up with this reality.

When you examine the organisation structure of most retail banks, the Head of Branch networks is second only to the Head of Retail, and in many cases, is a direct report to the CEO (Figure 2.1). In comparison, the manager responsible for the Internet often sits under the IT or marketing departments, three or four levels below the organisational equivalent of the branch business unit head. So let's get this straight. Ninety per cent of the transactions go through channels that are managed by managers who have only a modicum of influence within the organisation structure, while the head of branches has the ear of the CEO and looks after just five to ten per cent of the daily traffic within the bank.

"Banking is necessary, but banks are not." Bill Gates, Microsoft Chairman, 1994

"Ah, but the branch generates all the revenue…" we've heard it argued. This is a really good justification for keeping traditional structures in place. Well, let us examine if that is really the case.

Let's take credit card acquisitions as an example. How do we market credit cards? Currently we might use direct mail, newspaper advertisements, Web and possibly promotional marketing offering a "free gift" if clients sign up for a new Visa card or Mastercard. Customers are then faced with probably two or three choices of how to apply. The first option is that they can phone the call centre, but the call centre refers them to the branch because they need to present proof of income and proof of identity to an officer of the bank. The same might be the case for the Internet, where the application can be filled online, but we then call them and ask them to come into the branch to complete the application.

Who gets to record the revenue for the credit card application? Not the call centre, or the Internet channel, but the physical branch that executes the final signature on the application form and the Know-Your-Customer compliance check on the proof of income—they get to record the revenue of the sale. But the branch actually has had practically zero involvement in the sale, and is simply just a "step" in a required adherence to an outmoded

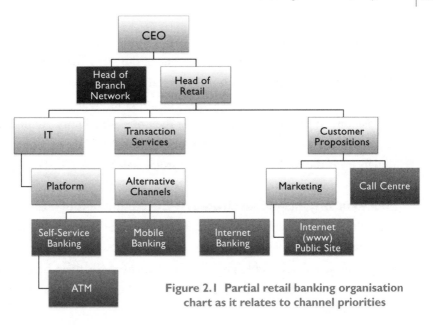

Figure 2.1 Partial retail banking organisation chart as it relates to channel priorities

compliance process. So does the branch actually generate the revenue, or does it merely handle the accounting?

> Branch revenue does not equal branch sales results. The decision to "apply" for a credit card is almost never made in the branch, unless it is a new account opening process. The branch has practically zero involvement in credit card sales, but we often require customers to come to the branch to actually submit the application. The branch is simply a "step" in an outmoded compliance process. So does the branch actually generate the application, or does it merely handle the accounting? Branch bankers would have you believe that the recognition of the sale should go to the branch when in reality, the whole compliance process just costs the institution money, the consumer time, and was not the best way to handle the application in any case.

The attitude of many retail banking senior executives seems to be that the branch is a serious banking channel, whereas the remaining "alternative" channels are just that—alternatives to the "real thing". The problem is that customers simply don't think like this. They don't assign a higher value or priority to the branch; they just see it as one of the many channels they can choose to do their banking. In fact, many customers these days choose not to go to the branch because they don't want to stand in line, or they find it troublesome to get to the branch during their operating hours. Admittedly the branch is the premium service channel, but it is NOT the only channel. So why don't the banks think the way customers do?

Policy and organisational strategy—the bailout fallout

One of the key factors that obfuscate the ability of the organisation to serve customers is the increasing adoption of bank policy in the name of risk mitigation or reduction. No greater an opportunity has there been to see this conflict of organisation and purposes than during the global financial crisis of 2008–09.

The issue of the bailout funds was hotly contested and argued in the US, UK, EU, Australia and elsewhere as very expensive mechanisms for preventing a 1929 type global depression. There were consistently two arguments given for injecting capital into the ailing banking system. The first was that the asset-backed securities underlying the subprime bubble had become "toxic" and only by purchasing these toxic assets could the market come to terms with the ongoing factoring in of these assets. The second was that the crisis had created a liquidity and capital adequacy crisis for banks and that they could only free up funds for the general public if their liquidity was improved.

The first goal may have been accomplished, although the long tail may yet still appear. The second goal was a failure in respect of customer expectations, however.

While banks achieved a welcome top-up that reduced their cash flow problems, internal risk strategy dictated that in an economy in trouble, all but the very best customers represented too great a credit risk to chance lending them money. So banks started to freeze loan books, aggressively pursue marginal accounts that were having problems meeting their repayment schedules, and basically stopped all lending to those that needed it—small businesses and individuals. Small business activity and retail consumption are two critical levers in kick-starting an economy after a recession, thus the bank policy on credit adversely affected the recovery cycle. In the meantime, as regulators got tougher on banks and investment firms, institutions sought to maximise fee and margin on lending products out of fear that regulation would restrict future options in this regard.

There was another significant factor here though. In taking government money, many banks suddenly found themselves, within the space of a few months, cashed up. As they weren't lending to customers, what would they do with all this money? They invested it, of course. Based on the Warren Buffet school of successful investing, we all know that *reversion to the mean* was guaranteed to restore value to the markets once economic figures started to turn around, if only back to the historical averages. Most consumers weren't that confident, but the bankers know a good trading bet when they see one.

Margin trading off government bailout funds, basically free money, created some very healthy returns in the space of just six to nine months. So instead of using bailout cash to bolster lending to those that most needed it, banks used the funds to generate profit for the bank. Now, if this resulted in more dividends for shareholders and the relaxation of some of the bank policies on lending for consumers, then this would be a fair result. But, instead, bankers decided that as they had all done so well, they deserved a hefty bonus for their hard work and clever financial mathematics.

Customers understandably were and are not impressed by such a brutal, net margin-led approach to policy and bank strategy. They rightly expect the bank to act as a service organisation and to look at even more opportunities to provide support for customers during a period of economic instability. That was not to be the way for the likes of Bank of America, JP Morgan, Citibank and others. The customer response was overwhelming and fierce.

Since 2008, it is estimated that blogs extolling the negatives of banking have increased by more than 400 per cent. A specific campaign in the US first covered by *Huffington Post* and then supported by major networks such as ABC encouraged consumers to move their money out of big banks and into credit unions.[4] Customers angered by opportunistic credit card rate hikes and overdraft fees flocked to YouTube to tell others of their treatment. Barack Obama, Gordon Brown and other leaders criticised the "fat cat" bankers and their bonus schemes, even slapping heavy tax penalties on future windfall gains. Some bankers reacted with anger that they should be questioned in this manner; after all "They need us more than we need them", the bankers say. Well, the momentum of the customer movement behind this issue may prove exactly the opposite—the bankers need customers on their side more than ever.

Customers are seeking alternatives to big banking because they believe that banks have lost touch with reality and don't care about their customers anymore ... and they'd be right. Let the banks cut back on the bonuses this year, and invest in reaching customers and improving their lives.

Customer value innovation

An emerging field in customer experience and behavioural research and marketing is the area of value innovation. Value innovation, in strategic terms, is the creation of a superior customer value with a view to gaining a competitive advantage and/or rejuvenating the institution and organisation.[5] Whereas organisations such as Google thrive on constant innovation, traditional organisations such as established financial institutions find this more difficult to manage.

Innovation of the customer experience, however, is no longer a choice, but a necessity, and more importantly, a competitive weapon. As customer behaviour continues to evolve more and more rapidly, a culture of continuous improvement of the customer experience is required (Figure 2.2). This culture of innovation can then influence the entire institution, from management through to frontline staff.

This notion of value innovation[6] goes beyond changes in product, process and services, and includes new ways of:

- servicing customers;
- offering value propositions to customers;
- collaborating with customers;
- working; and
- networking competencies and resources.

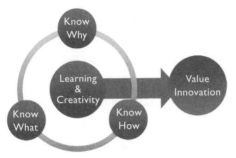

Figure 2.2 The customer value
innovation process

To measure innovation in the institution effectively, we need to measure how innovation is conceptualised, how well the innovation is oriented towards the customer (instead of solely internally), the approach that will be used to roll out the innovation or proof-of-concept, how we are learning as an organisation, and what research (R&D) is being carried out.

A number of institutions have recently appointed a Head of Innovation or Chief Innovation Officer to set out to accomplish this task. However, the major issue currently facing the institution is we simply don't measure collectively the experience of the customer, thus any attempts at innovation are likely to be met with frustration because the organisation does not understand the competitive benefit likely to be gained.

Why not try Google's approach? Google staff set aside 20 per cent of their working week to focus on innovation and improvement of the business. They can work on any project or assignment they like, and this often results in various project groups being spontaneously created by like-minded colleagues with an "idea" on how to do something better. Why not allow every department and product team to create an innovation role? Give the Chief Innovation Officer (in addition to the department heads) a dotted line of oversight of these resources, and let them go for it!

Customers? Who Are They Again?

The challenge intellectually for bankers undergoing this exercise is that while they do acknowledge they compete in an open market, they largely restrict their solutions to either product or marketing in scope. Indeed, pick up any strategy publication aimed at bankers these days and you're likely to read a great deal about database marketing, segmentation, customer intelligence and the like. The primary issue with this is that there has never been a time when customers have had more direct contact with institutions than they have today, and yet generally, institutional knowledge of how that contact can be managed successfully is very poor. We're not talking about customer contact management here; we're talking about how customers use the channels day to day.

It would make sense that certain customers at certain times of the day like to utilise a particular channel. It would also not be unusual to find statistically that certain types of products are better suited towards a specific channel because of either their complexity, the level of involvement required by the customer and/or an advisor or specialist, and other factors. So how well do institutions know all this? Generally, in our experience most don't have much of a clue because each channel is only measured in isolation. There is no conceptualisation that channels are related to one another because the organisation structure reinforces a one-channel-at-a-time approach to strategy.

Certainly most banks could tell you which products they've sold through the branches this month, and they likely track which products they sell online (if they have that functionality) or via the phone for revenue purposes, but rarely do they think about why. If I asked the average retail banker to tell me which are the best products to sell through their Internet channel, what would the basis of that answer be? If I were a banker looking at how to generate new revenues through my IVR or ATM channels, how would I figure out the right sales pitch and implementation? This is where things really break down in the average financial services provider.

There are two critical questions here, which we will delve into in much greater detail later. The first question really is, if the institution puts a product on a specific channel, will the customer engage them on that channel for that product? The second is, what is the best method to implement the solution so we get maximum take-up or utilisation? To be very specific, I'm not talking about origination, segmentation or marketing-mix here, I'm talking about which product or service works on which channel, how simple it is to access for a customer and how we measure that.

HOT TIP: Creating an Innovation Network
Giving employees the power to innovate

An innovation group or network across your organisation can be an extremely useful tool to generate new ideas, but unless this group is given some teeth and has some ability to effect change, then it is a waste of time. Here are some tips to making an innovation team work.

1. Seed the innovation team with key players from every part of the retail bank, not just full-time "innovation" employees. Every product team and department must participate.
2. Give each member of the team 20 per cent of their time to work specifically on innovation.
3. Include innovation "revenue" as a KPI (Key Performance Indicator) for each department head so that they will give their support to this team.
4. Create a blog or internal social network group for employees to suggest innovation ideas. Put it in the area where call centre and branch staff discuss problems that clients are having so that innovation is connected with solving clients' problems and issues.
5. The innovation group needs to present innovative concepts for at least one hour a month to the head of retail or CEO.
6. The CEO should allocate a portion of budget for innovation proof-of-concepts to be incubated. Those that successfully show revenue capability should be converted to a budget line item project.

Which product works best on which channel?

Let's begin by making certain assumptions that relate to the complexity of the product. Obviously the more complex a product, the more handholding the customer might need to engage in that product. By law in some jurisdictions, certain investment products, for example, require a customer to be advised as to the risk involved in that type of investment product. In other instances, a customer may have a plethora of choices and simply not know the right product for their particular circumstances, say, in the case of life insurance. A mortgage product is generally considered a pretty complex product (although increasingly commoditised), so it is a reasonable assumption that at some point before a customer receives approval, he's going to have to talk to a member of the mortgage product team to decide on the right option(s).

So investment products, life insurance and mortgages all require the branch, right? Well, it's not as straightforward as that. As a customer, I might use the Internet to research my investment options, so before I go to the advisor in the branch, or he comes to see me at my office, I may already have decided the asset classes I want to invest in, the investment horizon, the level of risk I am prepared to take. I may have gone online and used a risk profile questionnaire to see what level of risk tolerance I have. I may have used websites or magazines on investments to look at whether it is the right time to invest in my local property market or blue chip banking stocks. I may be part of an investment club online; I may even have my own online brokerage account separate from my retail bank. So while I may engage with an advisor in the final stage to execute a transaction, I may have already made the critical purchase decision preceding my meeting with the human advisor.

So while we can assume that execution of a complex product will likely occur through a face-to-face interaction, we cannot assume that this is the sum total of contact the customer will have had regarding that product. The fact is, for different products, there will be a range of different channel interactions that lead to the ultimate sale, application or trade. Understanding each of those interactions is just as critical as the face-to-face capability at the very end of the process (see Table 2.1).

So, we have to understand which channels are used by which customers at what time, and at which stage of the purchase cycle or decision they are at. The fact is, we really need to cater for all of these interactions simultaneously. The good news is that for simpler products it is somewhat more predictable. So, which products work best online? The answer is, it depends on which market and demographic you are looking at.

Table 2.1 Indicated channel interaction mix for investment grade products[7]

CHANNEL	GATHER INFORMATION	ANALYSE & EVALUATE OPTIONS	DEVELOP PLAN	EXECUTE PLAN	MONITOR & SERVICE
Phone	●	●	●	●	●
Face-to-face	●	●	●	●	●
Internet	●				●
Email		●		●	●
Snail mail			●	●	●
Branch/office	●				
Other	●	●	●		●

Here are some fascinating results from a 2007 and 2009 survey on Internet usage, where customers were asked the likelihood of whether they would purchase or apply for the following products online in the succeeding 12 months. The results below (Figure 2.3)—aggregated from 45 countries globally—show only those products where 40 per cent or more of those surveyed indicated that they would be *likely* or *very likely* to use the online channel to purchase or apply for that product type in the coming 12 months.

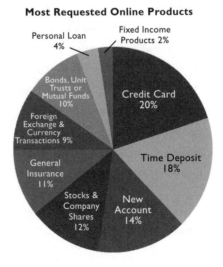

Most Requested Online Products

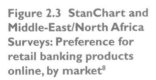

Figure 2.3 StanChart and Middle-East/North Africa Surveys: Preference for retail banking products online, by market[8]

We do see some similarities here. With the exclusion of the investment and trading products, all of the other products are pretty simple, namely credit cards, general insurance products, personal loans, time deposits or fixed income products, and opening a new bank account. These are also things we know work through the online channel. So how can we find this stuff out? Well, asking our customers would be a good start.

How do we ensure it works when we implement it?

A graduate stepping into a bank strategy meeting for the first time might be forgiven for thinking that he has an entirely new language to learn before being considered competent. The language of bankers often needs its own dictionary with terms such as draft (not a first edition of a document), telegraphic transfers (even though a telegraph is not involved), instructions (but not like an operating manual), annuity (what?), or let's try some acronym soup…

The language of bankers takes some time to learn. The concern here is that banks present barriers operationally just because of language. Such language may be what banks are comfortable with internally, but it is a language that customers may at best be puzzled over, and at worst, be completely disoriented by. Let me illustrate with a simple example.

Recently I was involved in a multichannel strategy assignment for a Middle-Eastern bank. As part of the assignment, we conducted a formal audit of all of the customer touchpoints, including the call centre and Web. During our audit of the call centre, we found an unusual anomaly whereby not an insignificant number of customers were ringing the bank because they were having trouble transferring their money to overseas accounts. The calls would go something like this:

CSR: Welcome to [the] bank. How can I help you today?
Customer: I'm having some trouble transferring my funds to my overseas account in London using the Internet banking facility. Can you help me?
CSR: Certainly, sir. Can you explain what the problem is?

Customer: *When I go into Internet banking and choose transfers, the list of banks does not show my bank in London. Can I add it to the list?*

CSR: *You should be able to type in the name of the bank and the SWIFT code directly onto the page, sir. What menu option are you choosing?*

Customer: *Third-party transfers...*

CSR: *Ah ... that's the problem. You need to be selecting telegraphic transfers. That is the option for overseas transfers. Third-party transfers are the option for transfers within Qatar.*

Ok, seems logical, right? Only if you are a banker. If you are a customer, it seems just as logical to assume that a third-party transfer means you can transfer outside of the bank to a third party, in this case, to your joint-account with your wife in a London bank. At this point, the bank really has two options. Either train the customers to understand the difference between a third-party transfer and a telegraphic transfer (at some very considerable expense), or use the language that is appropriate and logical for customers the first time around. Think like a customer.

What about if the bank simply changed the menu options within the Internet banking site to read "Transfers within Qatar" and "Transfers Overseas"? Is there any reason that the bank absolutely needs to use the most precise bank-centric language to describe these functions? If the institution designed the channel internally without any involvement of their customers, they would simply never know that this could be an issue because internally everyone understands what is meant when they use those terms.

This is a good lesson then. Plan on getting the customers involved as early as possible in the design process for a new channel or implementation of a channel. This process is known as **human interaction design** or **usability engineering**—making the channel usable, simple to use and relevant to the users who will use it. The cost implications of not getting customers involved as early as possible in the design process are extremely negative, and while most institutions baulk at the upfront expenditure, the resultant losses due to poor design are much, much more expensive than getting it right in the first place.[9]

> "Since the Internet is very much a part of our lives today, consumers often conduct online research before making a purchase. Of the 500 adults surveyed in the UK, 88 per cent reported using the Internet for research while 94 per cent use it for shopping. This means that even for companies which aren't selling online, the Web experience they offer visitors is key. Consumers visit a number of websites to find information and research a product or service before they make a purchase. This presents a golden opportunity for businesses to deliver a compelling and memorable interaction which will positively influence buying behaviour. By contrast, a negative online experience will elevate 'Web Stress' levels causing customers to click away…"[10]

> "The rule of thumb in many usability-aware organisations is that the cost-benefit ratio for usability is $1:$10–$100. Once a system is in development, correcting a problem costs 10 times as much as fixing the same problem in design. If the system has been released, it costs 100 times as much relative to fixing in design."[11]

So if you are a bank, how would you go about getting customers involved in the process? The initial involvement may be through focus groups or one-on-one interviews with key customers about what they need, but many banks are already doing this. **The better approach would be to get them actually involved in the design process.** Get them trying out early "lo-fi" prototypes of the interfaces and menu structures (IVR, Web, ATM), or get them to use the existing channels and observe them using those to evaluate the design problems before you embark on the redesign. With usability tests and observational field studies, we can normally identify 80 per cent of the critical problems with just five customers involved in the testing process. That is hardly going to break the bank, as they say.

Conclusion

Customer experience is no longer the sole domain of the branch. It exudes in everything we do, and customers are demanding a better experience across the bank. Inconsistencies in organisation structure, service levels between channels and silos frustrate customers who just want to deal in the most efficient way with the "bank". As we will discuss in Part Three, banks need to do the following now to build customer experience:

1. Appoint a customer champion to manage ALL channels.
2. Build analytics that identify failures in each channel, or touch points.
3. Get customers involved in the design process, particularly in respect of the interface and language.
4. Create an innovation team that has real teeth and give staff from every department time to focus on innovation.
5. Understand the total relationship the customer has with the bank across every channel.

KEY LESSONS

Customer experience is the new Holy Grail of retail financial services, but the key lessons are not so much about presence and service, as they are about understanding the core needs of the customer. To build an optimal customer experience you must be able to measure how customer behaviour is adapting, and be able to measure the rate of innovation or value creation within the institution.

While many institutions have the capability to measure customer experience, it is considered inferior from a data-set perspective to revenue. But whereas revenue tells you what you achieved, behavioural analytics can tell you why, and more importantly, where you can do better.

Keywords Customer Experience, Behaviour, Channel Silo, Cost Reduction, Organisation Structure, Branch Revenue, Product, Value Innovation

Endnotes

1   My apologies to Clarkson and May of BBC's *Top Gear*™

2   David Bell, Australian Bankers Association, quoted in ABC's
    7:30 Report

3   As reported in the *South China Morning Post*. 22 February 2007

4   See MoveYourMoney.info

5   Berghman, L., & P. Matthyssens, P. and K. Vandenbempt. "Building
    competencies for new customer value creation: An exploratory study".
    Industrial Marketing Management, Vol. 35, No. 8 (2006), pp. 961–73.

6   See also *Value Innovation in B2B* (2009), Cristina Mele, University of Naples
    "Federico II" Italy

7   UserStrategy.com

8   Standard Chartered Global Customer Survey – Internet Presence,
    April 2007

9   "The ROI of Usability", Usability in the Real World – Usability Professionals
    Association (www.upassoc.org)

10 "CA Web Stress Index", 2009

11 T Gilb, *Principles of Software Engineering Management*, Ormerudsveien,
    Kolbotn, Norway: Addison-Wesley Longman, 1988

Part 02

Fixing the Broken Bits— Channel Improvement

3 Rebuilding the Branch One Customer at a Time

TAKE a quick straw poll right now in the coffee shop you are sitting in, or around the office you are working in. Ask all of those sitting around you when was the last time they chose to go into a branch? Go on ... stand up and shout the question out over the jazz CD piping in the background ... just kidding. But it is a very fundamental question facing banks today: Do customers actually want to use the branch to interact?

If you asked most individuals when they had last been to the branch, the bankers amongst us might have us believe that the answer is going to be a jubilant "last week" or in the last few days—I love my branch! Most bankers I speak to still fundamentally believe that the branch is at the centre of the banking experience today. But just about everyone else you ask will probably tell you two things: "I can't remember the last time I chose to go into a branch" and "I actually don't want to go into a branch anymore; the only time I do is when the bank insists I do."

In recent surveys, 16 per cent of Australians said they NEVER go to a branch. Seventy-five per cent of Standard Chartered's customers surveyed online in over 40 countries said the Internet was their first choice of channel; only 12 per cent chose branch as their primary contact point.

Yes, bankers are on crack. Or if they aren't, they soon will be. The traditional role of the branch in our day-to-day banking experience is dying like George Bush's legacy or like Gordon Brown's personality. The backlash against "traditional" banking as a result of the 2008-09 global financial crisis is severe, intense and passionate. Taxpayers, who are seeing the leaders of banks (who just months ago were "rescued" by bailout funds) taking massive bonuses and patting one another on the back, are fuming.

Consumers of retail banking are standing up, blogging and tweeting, and demanding that banks do better and provide a more relevant service at a realistic price. As one frustrated Bank of America customer who had faithfully paid her credit card off for years without incident, commented: "Put that in your bailout pipe and smoke it!" Why? BofA had, without explanation, increased her credit card interest payments by 250 per cent in just a few short months[1] without notice, agreement or reason. Consumers are rebelling against what they see as a lack of justification for banking fees, compared with the value they are getting back from the bank.[2]

Such a backlash is not new. As we discussed briefly in the previous chapters, in the mid to late 90s in the Australian and UK markets, retail banks were finding that a large number of county/country, regional and smaller suburban/borough branches that served smaller communities were not profitable. The result was that the banks started to close large swathes of such branches to reduce costs and increase net earnings to impress shareholders. Bank valuations at the time benefited marginally from a cost cutting structure that showed fiduciary prudence, but large groups of disaffected customers launched protests and complaints at the big banks.

In Australia, the responses across the community and in the public domain were damning, especially when banks started to announce record profits as a result of the closures. Between the Big Four banks in Australia, the *Reserve Bank of Australia Bulletin* reported that 557 branches were closed during the period of 1993–96.[3] This trend continued well into the first half of the new decade.

In the US in recent times, BankVue and their Kasasa banking unit have built an extremely successful model as an integrator for smaller banks. They have taken marginal players—smaller banks from communities around the US—and given them a platform for greater customer involvement and profitability. Kasasa already has 15 banks enrolled onto its common platform. It gives customers of those banks a much broader service network because they can go into any Kasasa partner bank, use their ATM, process a cheque over the counter, or pay bills. It suddenly gives a smaller community bank the competitive capability of a much larger brand in respect of network and services.

The key to understanding the success of Kasasa and revisited community banks such as Bendigo Bank in Australia is realising it was not the branch as a channel that was broken, but the business model applied by traditional bankers to measuring the channel and measuring successful customer engagement.

In the UK and US, a similar experience was occurring with the consolidation of the post-deregulation financial services market. Banks were increasingly coming under attack for branch closures. In the UK, 3,600 bank branches were closed between 1990 and 1999, with more than 3,000 additional closures estimated for the period 2000–05.[4]

In the US, brands such as Bank of America, WaMu and Citi were increasingly in the news between 1996 and 2000 as a result of mergers and branch closures. In 1998, WaMu closed 161 branches in California alone. Many of the banks found they had overdone it in respect of these closures, and a resurgence in branch banking has been seen in recent years.[5]

Indeed, we would advocate that we do need more customer contact points, but not just the same old branch design. Banks need to innovate around form and function to get a better fit with customer needs. This has to be based on needs of the customer in respect of product and service, and not a transaction platform as traditionally held.

This uncertainty around branch function, cost structure and capability serves to illustrate the key differences in goals between the institution and the customer. The institution sees the branch increasingly as a revenue centre whereas in the past these were more accurately classified as cost centres. Thus, improvement in branch profitability has largely been the focus of the institution over the last 20 years.

Customers, on the other hand, expect service from a branch, and they expect this because they "pay for it" with account-keeping fees, over-the-counter fees and other such levies. For an exchange of "value" to occur between the institution and the customer, both parties need to be getting something out of this real estate.

What is the value exchange platform for the branch today?

The Core Function of the Branch in the 21st Century

To evaluate the role of the branch correctly in the modern 21st century retail bank, we need to consider what function the branch holds today. Most bankers would typify the majority of their branch network as "one-stop shops" for retail customers—a place where you can walk in and get the answer to any banking question your heart desires, apply for any product, or execute any transaction. The reality is, however, that most customers visit the branch increasingly less for day-to-day banking interactions. Let's examine the trends in some of the key markets already discussed.

United States and Canada

January 2006. In a multichannel world, the traditional bank branch continues to lose ground as a method of banking, according to an annual marketing research study by TNS Canadian Facts. Conducted in the fall of 2005, the firm's "How Canadians Bank" study found that just over half of Canadians (54 per cent) reported visiting a branch in the month prior to the survey, the lowest level of branch banking reported since the tracking study began in 1994. Not only are fewer Canadians visiting a branch, but also the number of monthly visits for those who do go to a branch has been steadily declining.[6]

April 2009. According to a 2009 survey conducted by Celent, 75 per cent of respondents have "definitive plans to increase branch count in the future", with only 3 per cent saying they have plans to reduce geographic branch footprint.[7] In branches, 30 per cent of retail banks surveyed reported a decline in transactions. Among mid-tier banks (assets of between $1 billion to $50 billion) surveyed in the study, 43 per cent cited a decline in branch activity and transaction volume, yet many of these still have plans to build more branches. The bulk of transaction declines are linked to teller transactions, where 90 per cent involve cheques.[8]

July 2009. More than two million US households have adopted online banking and bill payment during the last year, meaning the services are now used in over three quarters of homes with Internet access, according to a Fiserv-sponsored survey. The poll of 3,029 households—conducted by Harris Interactive and The Marketing Workshop—indicates that 69.7 million households now use online banking services, primarily to access balance and account history and transfer money. In addition, approximately 64.4 million households pay at least one bill online, either at a bank site or directly at a company portal.[9]

United Kingdom

September 2008. It has been a valiant struggle, but the odds are stacking up against the survival of the British high street as we know it. Reeling from a spending reverse by cash-strapped shoppers, town centre retailers are now facing a growing and powerful threat from out-of-town shopping developments. And as Lloyds TSB prepares to close up to 1,000 bank branches after its acquisition of HBOS, experts fear that an unprecedented number of empty high street units will soon disfigure market towns across the country. The rapid growth of the Internet, and sharp increases in rents and rates have placed added pressure on operators.[10]

Australia, New Zealand

April 2007. Nielsen in April of 2007 reported that the trend of moving away from branch to Internet banking and ATM was increasing. Forty-nine per cent of customers said that they visited the branch less than once a month, and 16 per cent said they never visited a branch.[11]

September 2008. ANZ Bank called for voluntary redundancies among its branch staff and announced a review of the branch network with a view to reducing staff numbers; a freeze on recruitment which will lead to immediate staff reductions through attrition; a reduction in the use of casual staff to cover absences; and a tightening up of overtime being worked. In explaining the reasons behind the move, documents obtained by interest.co.nz detail how over-the-counter transactions have fallen 23 per cent to around three million per month since 2003 because of simpler processes, fewer forms, more use of call centres and more use of Internet banking. However, branch staff numbers had been stuck at around

1,590, as the documents show. Wayne Besant, managing director of ANZ retail banking, issued a statement on the plan.

"Today we announced to staff in our ANZ branches that we will be conducting a review of customer activity within our branches. Our goal is to align our service model with customer demand,. The business of retail banking is changing. Customers are utilising and adopting alternative and more convenient ways to conduct their banking transactions, such as over the Internet or by phone. As a result, and in line with general industry trends, we're seeing a continued decline in the volume of over-the-counter transactions in our branches. Branches were, and always would be, a very important channel for ANZ, enabling face-face-contact with customers. Maintaining excellent customer service will always be our priority."[12]

Europe

January 2006. "Customers can do everything using the Internet," explained Diederick de Buck, systems programmer and technical architect at Rabobank. "They can transfer funds, make payments, apply for a mortgage or personal loan, execute stock transactions, and more—and with over 1.5 million unique customers every month, we have the largest Internet banking site in Europe. When you combine Rabobank's Internet banking with our extensive ATM network, it's not surprising that we have seen a 90 per cent decline in branch office visits over the past three years."[13]

March 2009. The NCR Self-Service Consumer survey, conducted by BuzzBack Market Research, reveals that 67 per cent, or nearly seven out of ten consumers across France, Germany, Italy, Spain and the United Kingdom are more likely to do business with companies that offer the flexibility to interact using self-service, whether the channel be mobile, the Internet, or through an ATM or kiosk. Additionally, 56 per cent said their likelihood of using self-service has increased over the past year. Bill Nuti, NCR's chairman and chief executive, delivered the results during the company's Self-Service Universe conference in Budapest:

"The bottom line is that we are truly at an inflection point. Technology innovation, coupled with changes in consumer behavior, is forcing businesses to adapt to a consumer who is changing the way they connect, interact and transact with your business. These survey results are symbolic of what lies ahead for self-service. The self-service revolution is real because consumers see how they personally gain from it."[14]

So change is inevitable—deal with it and thrive

The facts are that while traditional branch banking is under threat, it is not enough to attribute this to the increasing rise of Internet banking and ATMs or to classify this simply as the "death of the branch". Some bankers proclaim a *return to branch banking* in recent years as we realised we had overdone it with branch closures, but this is also an inaccurate reflection of the facts. **The reality is that the role of the branch must change in the future to survive.** As a result of technology adoption diffusion and pressures resultant from shifts in consumer behaviour, if we don't adapt the role of the branch, we will soon see the business model for such made completely unsustainable from a cost-benefit perspective.

The sooner retail bankers see the Internet, branch, ATM, phone and other such channels collectively as the "bank" rather than seeing the branch as a somehow more superior channel, and the sooner we understand that the role of the branch must evolve, then the sooner we can get on with really servicing customers appropriately.

If you take just the one element of cheques (or checks)—the fact is that a large portion of UK and US branch traffic has been attributed to cheque processing. Now, however, banks such as HSBC are seriously considering phasing out cheques over the next two to three years. If cheques were to disappear, how would this affect the operability of your branch? The fact is that much of what you do today would simply no longer be needed. This is happening. Get ready. In some markets we've observed, traditional branches are being decimated unless they significantly change their operational practices. Those that have maintained healthy branch activity in major centres have two elements in common, namely:

- They have shifted branches away from being transactional centres to service centres that have been optimised based on customer segments or service type.
- They have adopted non-traditional operating hours, such as opening till later at night and on Sundays when customers have time to visit branches.

As we will see later in this chapter, branches are already beginning to change in both form and function. The core function of the branch moving forward will be about establishing the relationship with the customer at inception, and extending that relationship through an advisory sales process and excellent customer support systems. It is conceivable that all of the transactional elements within a branch will be moved to automated banking within electronic banking centres, automated branches, ATMs or the Internet within the next 10 years. What then is left? The face-to-face, value-add of a real, live human interaction.

As we've seen with stock trading, a transaction "platform" has no value being situated in a physical branch because the human "teller" or "broker" generally offers no value add to that transaction. Indeed, very few traditional brokers have survived the Internet trading onslaught of the dot.com boom. If we are honest, the only processes which will truly require a face-to-face interaction in the branch in the future are those that are sales and service related. These are also the only elements that will continue to make branches viable from a cost-margin point of view as the current over-the-counter transactions will simply remain a cost, rather than revenue opportunity.

In regional centres, we've seen that community banking centres have thrived where big banks have failed. Why? Again local knowledge that translates into better service for local small businesses has been the key to reviving unprofitable small town branches, not the branch in itself. Knowledge of what the customer needs and streamlining the process to solve the needs and problems of customers is a true differentiation, not products or interest rates.

So what is the answer to the branch conundrum? Closures of branches are not necessarily the answer. Many of the traditional high street branches will inevitably close as decreasing traffic and increasing costs will increasingly make them no longer feasible in the current evolution. Flagship "brand-store" branches may emerge, but will need to change in function *away from* traditional high-counter, transactional focus to low-counter, sales and service focus. However, more than that, a "standard" type branch is no longer going to be possible in the **BANK** 2.0 paradigm. We need to adapt to where customers are going to be physically, what services they are going to need and how they want to work with the bank. That is going to mean a number of different types of branches filling key service niches in the marketplace, not just a cookie-cutter approach.

Branch Models Emerging
That Will Survive in the Bank 2.0 World

So the core function of the branch as we move forward has to be about providing exceptional service to existing customers, and about acquiring new customers and share-of-wallet through better sales capability. If that is the case, what does the branch look like?

The concepts discussed below are not futuristic branch concepts that might or might not come to fruition; they are successful adaptations of the branch concept that are working today in different parts of the world. While you might not choose one of these specific models, the core elements that make them successful are things you should choose to incorporate in your branches today.

Flagship branches—the virgin megastores of banking

In many ways, the high street branch has been as much about real estate as it has been about presence and capability. In a 1974 speech to a friend's MBA class, Ray Kroc, the founder of McDonalds, asked the audience what business they thought McDonalds was in. The answer appeared obvious— burgers and fries! Wrong. Ray Kroc explained that McDonalds was really about real estate. The question for McDonalds was really about where to position the store to maximise visibility, flow of business and future capital gain on the property. In the early days, they actually purchased the real estate and leased it back to franchisees. However, the financial success of an individual franchise often came down to where the restaurant was physically located. It is still an essential part of McDonald's planning of new franchises today.

In many ways, banks face the same decisions in respect of their high street presence today. Such branches are often very high cost because of the rental rates, but the brand exposure alone from the presence makes a certain amount of sense. So how should the institution utilise these high-cost locations effectively?

The objective of these branches must be to show the capability of the brand, and to get the sales which guarantee the margin that makes the branch financially viable. Cash transactions are henceforth relegated to an "automated banking" zone where ATMs, cash and cheque deposit machines can be easily used, perhaps with a meeter-greeter who directs transactional customers to those facilities. With the high-counter teller space removed, we are left with low-counter, coffee shop-style seating for specialist advisors and sales staff to engage customers in a friendly, brand positive environment. In addition, depending on the target segmentation, these "brand" stores could be built to attract specific segments, with HNWI (High Net Worth Individuals) being attracted to more of a luxury brand-style store, or younger customers to a coffee culture-style store.

Deutsche Bank decided in 2005 to create one such store in Berlin. The Q110 design is as unusual a branch as the name; Q110 stands for Quartier 110 on Friedrichstrasse in Berlin. Customers enter a branch without barriers or counters that keep visitors at a distance. In the Forum, customers meet relationship managers for their initial informal discussions. Kids Corner is where children are looked after professionally while their parents concentrate on banking matters. The Lounge is a place to relax in, to chat with friends, and enjoy tasty snacks and refreshments. The Trend Shop has a constantly changing assortment of new and trendy products. On top of that, customers in Q110, as in a supermarket, can find the financial products they are looking for in attractive product boxes that they can take home with them. Bank products become tangible.[15]

Deutsche Bank reports that the experiment of the Q110 branch was an extremely successful one, so much so that they plan to roll out similar branch concepts in Munich, Stuttgart and other locations within Germany. As Deutsche Bank manager Ira Holl, says:

> "To make a visit to the bank as simple as possible and a different experience: that is the philosophy of Q110, our new branch concept for the Deutsche Bank of the future. An innovative idea made real: a bank with no counters or barriers. The response from customers and staff has been very enthusiastic."[16]

Figure 3.1 Q110 branch concept from Deutsche Bank
(Credit: Deutsche Bank, under Creative Commons)

In the Danish market, Jyske Bank has introduced a new concept under the name, "Jyske Differences".[17] Products are presented physically as packages, and the bank's interior design works to improve customer interaction and service outcomes. Like Deutsche Bank, Jyske has also thought about re-engineering the physical space to make the experience more like visiting a retail store. Some highlights of their redesign include:

- calling the branch a "shop";
- using a hotel-style check-in desk as their transaction/teller station(s), which they call the MoneyBar;

- featuring a concierge, called the AskBar;
- calling the area between the various bars The Market Square, which is where you will find Theme Island, with stacks of financial products packaged in boxes;
- a TestBar, where you can scan any of the boxed products for an on-screen tutorial;
- an area they call Oasis, which looks more like a reading room than the waiting room it would be in a regular branch; and
- a prominent CoffeeBar for its customers to leverage the "coffee culture".

Wells Fargo has tried a similar retailing approach in the United States, moving to rename their branches "stores", remodelling branches to make them more appealing, and trying to change the branch culture to reflect a more service oriented approach for customers. As former CEO Dick Kovacevich said in respect of making branches more profitable and turning them into stores which customers feel happy to use and visit:

> "And I would just ask two rhetorical questions: Who over time have been the better merchandisers, retail stores or banks, in terms of their ability to attract customers and serve them well? Most people would say retailers have been more effective than banks. And I'd ask the second rhetorical question: How many retailers don't want customers in their stores?"[18]

Other significant changes included a complete transformation of the forms and documentation used within branches both to market to customers, and for the process of applications and so forth. For example, Wells Fargo used ethnographic research, auditing and analysis to reduce over 200 documents to six core brochures as part of their customer focus strategy.[19] More detailed information was moved to a series of online fact sheets. The result overall was that Wells Fargo cut millions of dollars in marketing costs and improved the clarity of their core messages to customers.

Here is a direct quote from Wells Fargo's annual report that shows the role technology can play to improve the ability of the bank to service the customer more effectively:

> "We use technology not to de-personalize service but to personalize it. Thanks to technology, we know how many products each customer has with us. We can anticipate the products they'll most likely need—based on account balances, life events, transaction history, and how they access Wells Fargo. With that knowledge, and respecting the confidentiality of information of our customers, we can tailor sales messages ... to help satisfy our customers' financial needs."[20]

What typifies such changes is almost always a significant shift in the service culture of the bank to move towards a feature and service-rich environment for the customer. Banks have realised that products no longer differentiate, brand is not the sole factor in success, and branches themselves hold no particular ability to draw customers, so the ability of the bank to meet the needs of the customer must drive new revenue opportunities. Branches become a platform for this, but only with significant re-engineering from a personnel, design and technology perspective.

Bank-shops and pop-up branches

We realised in the 90s that people weren't getting to many of our high street branches as much as they used to because of the demands of modern working life. It was becoming increasingly difficult to get time off work to "pop down to the bank" during the lunch hour, and if you did, you'd suddenly find yourself presented with a queue that looked like people were lining up for the very last Michael Jackson concert. Branches remained largely empty during much of the working week except for spurts of high-demand activity at lunchtime and just before the branch closed at the end of the day.

The obvious answer might appear to simply open branches a little later and extend the opening times late into the evening so that people coming home from work would still have time to get to the bank. However, security concerns, issues with staff unions, the bank culture as regards branch hours and other such complications made such progression difficult. Taking into consideration the primary objective of getting people into the branch, the question became where could we put a branch where we know people are going to be?

When you consider the habits of people in western societies, and increasingly in other cultures, evenings and weekends were often spent at shopping malls, cinemas and entertainment complexes, and downtown "hot spots" for restaurants, pubs and coffee shops. Thus, the bank-shop was created.

To some extent, because bank-shops weren't a "real" branch, we could get away with selling the concept of these nouveau branches to management as a new initiative. Bank hours had to be more flexible because it would look a bit strange having a closed bank-shop in a major shopping mall on Friday night when the mall saw its maximum traffic. Since then, in various locations, these bank-shops have been huge successes. Location and availability (opening times) are a key driver to success here.

The other advantage that bank-shops offer, apart from accessibility, is the ability to predict the service and product requirements of the branch traffic better. While in the high street branch we have everyone from students, to families, to HNWIs, to retiree/passbook holders frequenting the branch, in bank-shops we could typically narrow down the demographic and their requirements more precisely. While customers may enquire about a mortgage at a high street branch, it is more likely in a bank-shop that customers look for information on a credit card, high interest or fixed deposit account, or a car loan.

Taking the concept of the bank-shop to its logical conclusion, we find that there are perhaps many more opportune places where branches could be located either permanently or temporarily to maximise the sales or service opportunity presented by the locale and the potential audience. In reality, a branch doesn't even need to be a permanent structure. As long as there is a brand presence, qualified sales staff and the ability to interact with the bank systems to support the sale or transaction, it is enough.

Thus, as we move forward, we as bankers will find more ways to get access to customers in a more directed, purpose-built fashion that maximises the selling or service opportunity. Table 3.1 lists a few examples of pop-up branch configurations.

Table 3.1 Pop-up branch configurations in the BANK 2.0 paradigm

Branch Type	Location or Event	Product or Segment Specialisation
Branch stall	Trade show	Mortgage (real estate trade show), car loan (motor show), etc
Home shop	Display village or model apartment	Mortgage (real estate)
Auto branch	Car dealer	Car loan and leasing options
Preferred customer shop	Airport	HNWIs with local advisory requirement, foreign exchange desk
Ship-shop	Cruise ship	Investment advisory for retirees
Branch-in-a-truck	Popular weekend spots such as "markets"	Credit card services, personal loans, etc
Co-branded	High-end high street (Louis Vuitton, Mont Blanc, Chanel, etc)	Brand conscious HNWIs, luxury service centres
Briefcase branch	Lobby of large office complex or head office of corporate account	Workforce/group life and health, and tailored business services
University store	University campus	Student loans, etc
High-flyer	A380 Airbus (long-haul)	HNWI and corporate advisory for business travellers
Pop-up portable branch	Anywhere it's needed	Specialised segmentation or sales pitch for audience at target

Coffee shop, automated and self-service branches

We know from Michael's introduction that these types of branches have not always been an unmitigated success, but increasingly in the right locations, such automated branches are now flourishing. The trick here is that these should be mostly new branch locations, or should be seen as new value added branches, rather than replacements of "'personed" branches. If a human-staffed branch is closed only to be replaced by an "automated" branch, the perception will be a decrease in service. However, if a new, hi-tech, automated branch or a coffee shop-style branch opens in a new location, the bank is seen as innovative.

The likes of Rabobank, ING Direct and others have been increasingly successful with cashless, teller-less branches in the form of coffee shops. The only transaction you can execute with cash at these branches is, in fact, the purchase of a latte or cappuccino. The teller is non-existent and the access to bank services is either via an Internet terminal or via the ATM and deposit machine near the entrance. As a result, both ING Direct and Rabobank have more than 90 per cent of their daily transactions occur through the Internet, phone banking and ATM network.[21]

Now the purists would argue that ING Direct and Rabobank have been successful with "e-channels" only because that is the *only* option customers are provided with. Yet the fact that the segment of customers they target seems genuinely happy with the service and the bank's profits would indicate that they have not suffered due to a loss of branch function.

What about traditional banks? Well, HSBC has had huge success with FirstDirect in the UK, again starting with phone banking in the 80s and then increasing its focus on the Internet over the last decade. Even HSBC in Hong Kong has stated that 90 per cent of its daily transactions occur through electronic channels.[22] So it would appear that it doesn't matter whether you are a traditional bank or an Internet-only bank; the trend towards greater use of automated services is universal. So why shouldn't this apply to branches also?

The traditional players can, of course, take a slow and measured approach to this by creating hybrid branches with some automation for the transactional side of the business, but low-counter capability for sales opportunities. In Asia it is now very common in Singapore, Hong Kong and Shanghai to see meeters and greeters tackle an incoming customer to enquire "what are you looking for today?" If the answer is cheque deposit, withdrawal or transfer, they are directed to a "bank" of automated devices that can more than adequately serve their needs for those types of transactions, freeing up the floor for more profitable transactions. Likely the trend will be for even greater use of technology in branches as we move forward.

A simple improvement through the use of technology, for example, is HSBC's recent initiative with their premier customers in Asia. Trialled initially in Hong Kong, HSBC introduced RFID (Radio Frequency IDentification) cards for their mass affluent Premier customers. Adopting the same type of RFID technology used by retailers in shopping malls, the proof-of-concept demonstrated how the most valuable customers could be served better by integrating this technology. As soon as Premier customers entered the branch, they were recognised as a priority customer and the purpose of their visit was captured. This ensured that once the customer approached a Premier relationship manager or customer service officer, staff were already aware of the customer's details and reason for their visit. The teller doesn't have to go through the standard process of asking for their name, address or account number again. Of course, secure identification or a signature is still required for a third-party transaction or withdrawal, but the service perception for customers goes through the roof. On top of that, HSBC also offered free wireless at the premier branches.

The Y-Generation and the next generation of customers have grown up on Skype, Facebook, Twitter, YouTube, Instant Messaging, SMS and countless technologies that have made an art form of digital interaction. This generation would think absolutely nothing of walking into ABN Amro's teleportal branches. ABN Amro's Teleportal Multi-Access Bank-shops use only videoconference teller access, where one teller can simultaneously look after a number of branch locations, but all cash and cheque transactions are taken care of by devices in the foyer of the branch. Initially designed for university campuses, these so-called teleportal branches have been so effective that ABN Amro is rolling them out across Europe for many different types of segments.[23]

If we introduced the green movement or issues of the environment, we would increasingly be required to think about "paperless" branches that are more carbon-footprint efficient. Thus, the more automation we include and the less paper we use, the better. Therefore, our banking experience will

Figure 3.2 ABN AMRO's Teleportal Multi-Access bank-shop model
(Credit: ABN Amro)

be more integrated with the technologies available. For example, instead of printing off a receipt from an ATM, the ATM will automatically send our account balance and last five transactions to our mobile phone when we complete a transaction.

Increasingly, we'll find ways to use our NFC (Near Field Communication) mobile, our Oyster or Octopus card, or debit cards for micropayments, decreasing our need for cash in any case. When customers appear at the "counter", why ask them to sign their name on a piece of paper when we can get them to sign on a tablet, or present their fingerprint as a suitable unique identifier.

When we pay an electricity bill, our Internet-enabled fridge may give us the good news on its display, or our mobile phone will simply log our bill payments. Renewal notices for outstanding tax or utility bills will be sent to our mobile devices, or even our car, to provide us with reminders. The more we neglect such outstanding payments, the more persistent our devices might become. Alternatively, our smart bank account may be able to manage such day-to-day bills and transactions on our behalf. We're not talking direct debit here, but actually a paid-for banking service provided by the bank as a value-added service. Forget EBPP (Electronic Bill Payment and Presentment), let's talk *Intelligent* Bill Management solutions provided by the bank or our telco provider.

In fact, the greatest improvement in automation is not about technology nor about cool interfaces, but simply about anticipating the needs of customers. With better information on what our customers are doing on a day-to-day basis, which channels they use for which transactions, and what are their total product/service needs, we can use behavioural models or analytics to anticipate and service the needs of our customers better. Primarily this will make itself evident in better, intelligent sales systems where the system will give tellers an offer to present to the customer on completion of his or her primary transaction for that visit. Rather than try to pitch them a credit card when they already have three separate cards from our bank, why not offer them a packaged home and car insurance solution for the new car they've purchased because they are also a mortgage client? Or offer them an upgrade to their personal loan with a redraw because they've been such a good customer.

So, greater automation for better service and improved sales revenue is a given. The only question is how we as bankers can integrate this into our overall service platform at a reasonable cost. The first step is to improve integration of such technology into our existing branches, reducing the cost of zero-margin transactions such as withdrawals, deposits and transfers, along with better customer "service" initiatives. The second step is to evaluate the potential for fully automated branches when we want to expand our real estate footprint of physical locations. The concurrent development steps are to improve the usability and interface of our customer-facing technologies so that they are not inferior options, but just choices. Then automated services become just improved service, not only ways to reduce cost for the bank.

Third-party branches

The final type of branch evolution that we see emerging is the franchise, reseller or third-party branch. We've encouraged the development of brokers for the insurance industry and other elements of the financial services arena for many years, so why not branch banking itself? Community banks have been hugely successful in taking this approach by providing the platform and ensuring local involvement for success. However, as banking systems become simpler to use and more usable, and as we focus more on revenue generation over the counter, then we'll simply be able to outsource branches to franchise or third-party operators.

Now the immediate concern I can imagine hitting you in the face at this suggestion is the increasingly tough compliance requirements presented by regulators and lawmakers. However, in real terms we have systemised most of the compliance and risk mitigation processes today, and these are largely handled by the systems or processes embedded in the branch. With the right training and systems, this is not an issue as long as there is an effective quality control mechanism put in place to ensure ongoing standards are met.

Ideally, third-party branches will be purpose-built for specific sales activities or segments. The institution gets the benefit of having no real estate costs, and no staffing costs. The branch or "desk" needs to fund itself, so as a result becomes very revenue focused and streamlines all the non-essential activities we find in the traditional branch. Just like when you use a third-party ATM when you are overseas, you'll have the choice of doing a traditional transaction over the counter for a fee to the franchisee. If you don't want to pay that fee, then you elect not to use the third-party branch. Can it work? Well, the post office has been doing this sort of thing for decades, and the basic systems and service sophistication are quite similar. So why not for banking?

This has the added advantage of increasing brand-width rapidly, with very little outlay in terms of physical space or investment. The increasing "network" growth means capturing customers, increased revenue and fewer costs on the front line. But it does mean investing in systems that are largely foolproof operationally. The reality of this requires two very specific back-office technologies to be integrated into the banking systems, namely CRM (Credit Risk Management) and STP (Straight-Thru Processing). The roll-out of third-party branches is dependent on such technologies to ensure the franchisee or agent can adequately execute. The CRM systems reduce risk, and the STP guarantees immediate turnaround on a product application, rather than requiring the supplemental physical back office processes we still seem to be lumbered with today.

Incidentally, banks are going to require this investment in CRM and STP in any case to support the demands of customers who want faster turnaround, and want to know what credit facilities they qualify for without having to jump through the usual 1,000-point checklist, 35 years of audited financial reports and statements, mandatory DNA testing, and biometric registration.

In short, banks need to take a leaf out of the Starbucks concept of "branch" or store. A packaged, well supported, self-sustaining business model that is primarily revenue generating in its focus, but with fantastic service metrics. Take the low-counter handling of cash transactions out of the branch and this is entirely possible today, with the right investments in systems and process re-engineering.

Branch Improvements Today

So what is on the branch improvement roadmap that we can achieve in the short term that will bring both benefits to the organisation and to the customer? The following areas represent suggested opportunities for either improvement in financial operations or customer service levels at the branch over the coming three to five years:

- Improved customer communications and language.
- Better cross-sell/up-sell capability.
- Efficiency gains through process and training, including Rapid or Pre-emptive Credit Risk Assessment, Straight-Thru Processing of applications, and sales/service culture and training.
- Better channel migration.
- Improved use of transactional automation and service technology.
- Better segmentation and location management.

These improvements make themselves evident through a range of projects that can be undertaken within the branch. Some of these projects cross over the above areas of opportunity, so we'd like to list the projects as specific illustrations of how improvement and transformation are achievable. See Table 3.2.

Table 3.2 Improvement initiatives and desired outcomes

Project/Initiative	Desired Outcome
Cash/cheque deposit machines	Reduce OTC (over-the-counter) transactions that are purely cost for the branch.
Meeters/greeters	Redirect non-optimal transactions to self-service automated capability.
Customer information system	Improved behavioural analytics on customers across all channels to understand better which "tasks" customers prefer to do at the branch versus online, etc.
Sales intelligence and automated offer capability	Real time and pre-cognitive offer management for existing customers delivered in the form of prompts, offers, or service messages.
Branch customer dashboard	Customer information dashboard that shows entire relationship footprint at a glance, along with current risk rating, credit approvals and suggested sales offers.
Improved staff mobilisation	Focused service and sales training programmes, along with better KPIs that focus on more than simply the number of applications per month, or total revenue.
BPR (Business Process Re-engineering) on select processes	Reduction of layering between sales and service departments, including the removal of duplicate "skills" within "competing" product units. Creation of "customer dynamics" capability as owners of customers, rather than product competing for revenue from the same.
Straight-Thru Processing (STP) and Credit Risk Management (CRM) systems	Enabling customers to get immediate fulfilment for an application rather than waiting the obligatory 24, 48 or 72 hours due to antiquated manual or human "processes" in the back office. Results in improved service perception and reduction of abandonment due to ongoing process demands (i.e. proof of income, faxing of 3 months' bank statements, salary certificate, etc.). Additional benefits include reduction of compliance errors through manual mishandling.
Customer friendly language initiative	Use of ethnography, usability research, audits, customer-focused observational field studies and focus groups to improve language and simplicity of application forms and communications with customers within branch (and beyond).

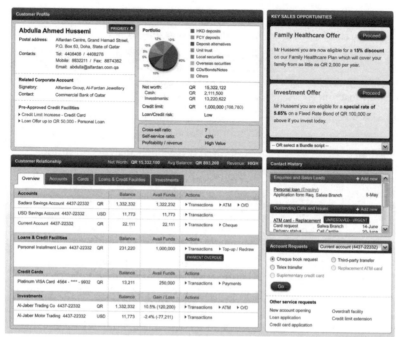

Figure 3.3 Proposed in-branch customer dashboard
(Source: UserStrategy, Heath Wallace)

Customer Dashboards
Branch staff need access to up-to-date customer details that enable them to form better, more profitable relationships. Key information should include total relationship footprint, sales opportunity (based on analytics and modelling), outstanding issues, and credit rating and pre-approvals.

It is not necessarily a requirement that all of these initiatives are completed at once, or even that all of them are completed. However, these steps are a specific formula for success. Each initiative is designed to save money and add value to both the bank and customers. They are also all initiatives that will eventually have to be done to keep the branch competitive and viable in the longer term.

For a modest regional or country "brand" bank, these initiatives should give a return on investment within the first 12–18 months. For larger multinational banks, the benefits of scale will reduce that time frame down to 4–6 months as long as the bank has a platform enabling knowledge transfer across the organisation and a technology platform that is integrated across the total network of the brand.

We believe in the branch. It's just that we believe in a pretty different branch to the type that is common today. The good news is, get this right and you'll not only save your most valued customers, you'll also make more profit for your shareholders. Just be sure to focus on the changing needs of the customer. Fewer transactions, more advice. The branch is a place I come into to talk to an expert on banking. I don't come into the branch to transact because there are so many easier ways of doing that.

If it is so easy, why hasn't it happened before this? It has. ING Direct, Deutsche Bank, Wells Fargo, HSBC and First Direct, Rabobank, and many others have got elements of this right already. It's just a matter of anticipating the changes in customer behaviour and moving the function of the branch to intercept that need.

KEY LESSONS

In the 90s, UK, US and Australian banks closed large swathes of branches in the name of cost cutting, creating large-scale customer dissatisfaction.

Recently branches have seen somewhat of a resurgence in popularity, but are branches working the way they need to work for both customers and banks in today's changing landscape?

This chapter reviewed what branches will become in the **BANK** 2.0 paradigm and how banks can make branches work much more effectively to create real value. It means that current configurations of branch are extremely unlikely to survive.

BANK 2.0 is all about less high-counter transactional support, and far more low-counter sales and service focus. The objective is to leave other less costly channels to handle no or low-margin transactions, and focus on where the value is—deep, profitable customer relationships.

Keywords Branch, Product, Future, Bank-Shop, Pop-Up Branch, Megastores, Automated Branches, Virtual Branches, High-Counter, Low-Counter, Sales, Service, In-Branch Systems, Dashboard

As a manager who learned his banking craft at Citibank in Australia, I had an unhealthy obsession with branches. Why? Well, at the time, Citi only had 24 of them up against our competitors which measured them in the thousands. We had an inferiority complex that dominated much of our decision making.

One of my first jobs was to assist in the roll-out of the model branch concept in the mid-1990s. The concept, similar to what Brett is talking about in this book, was to design branches to maximise the customer experience—from their interaction with the greeter (we had an automated queuing system), their ability to receive an instant ATM card, instant cheque books, all on the first meeting. The layout was designed to ensure that customers had to walk past the "advisors", who actually sold products, on the way to the teller. Somehow this was supposed to induce the customer to buy something on the way out or in; I'm not sure it ever really worked … but I loved this job! I had to pour over floor plans, spend lots of time with vendors pondering the latest MICR (Magnetic Ink Character Recognition) encoding equipment and pontificating as a zealot on "customer experience".

The first "model" branch opened in Adelaide. The branch manager was so excited as were we in head office—the result, well, we had a state-of-the-art branch that looked great, and we still had the same customers coming in. Did it reach our expectations? Well, I don't know what we expected to happen. When our product people had to absorb the cost of the new branch, then the reality set in. Apart from a couple of other "model" branches, the concept quietly faded away.

This was part of a bigger continuing headache for Citibank in Australia at that time. We had no branches, so how should we compete? A vast majority of our transactions were done through non-branch channels, yet research told us our prospective customers didn't use us because we lacked a big branch network. So what did we do? We opened up a token number of branches for "branding" and then spent a whole lot more on model branches to create a unique customer experience for a special few.

I felt we compensated quite well for not having a network. We had a large mobile sales force, a well connected introducer network made up of accountants, financial planners who sold our products, and we were part of one of the largest ATM networks in the country. Yet management always felt, "If only we had the branches that the big guys have."

Then an earthquake hit us. The arrival of mono-line financial providers into the Australian market in the early 90s. They came selling one product, at least initially, and they had no balance sheet, no branch network, yet they took a large share of the mortgage market within two years. And this was years before the Internet!

That's why when people talk about the "state-of-the-art" branch and get all excited about branch design, those groovy new seats, the big plasma TVs, calling it a shop, a spa, but definitely not a branch, well I say "been there, done that" 15 years ago. For me it does not depend on the branch; it depends on what the branch can do for the customer, what services it provides, if it is the most convenient choice, and what other alternatives I have that are more convenient.

So, my lesson to take away from all of this is this: it's not about a branch per se. It's about your business proposition, how you are going to sell and how you are going to service. The branch models Brett has identified all have something in common, that is, they are about changing the way the bank can effectively sell and service customers.

Today, we know that we can sell products remotely and we can also service remotely, that is, without a branch. We also know that for banks most transactions are done through non-branch channels, and increasingly sales are being conducted through direct sales channels. And for the people who say only humans can build relationships with other humans, I agree, and for high value relationships that justify a dedicated relationship manager, they can conduct that relationship anywhere, not just sitting in a branch.

So the one lesson bankers need to take away from this chapter is that your branch is not sacred. It is just a channel and your customers may choose an alternate channel to work with you. They may even prefer an alternative channel. Don't penalise them for that. Use it to your advantage.

Endnotes

1   "Debtors' Revolt: Woman refuses to pay of BOA credit card", *Huffington Post*, http://www.huffingtonpost.com/2009/09/14/debtors-revolt-woman-refu_n_285394.html

2   For discussion, see "The Customer Value Exchange" in Chapter 5

3   "Bank Branch Trends in Australia and Overseas", *Reserve Bank of Australia Bullentin*, November 1996

4   Bank of England, see also: "Bank closures cause concern", *BBC* Website, 6 September 1999

5   "The Branch is Back: Global Case Studies in 21st Century Banking Success", David J. Cavell, May 2008

6   "Visits to Bank Branches Continue to Slide": STUDY, Migration to Electronic Channels Continues, TORONTO, 31 January 2006, TNS (www.tns-global.com)

7   "Gathering Deposits Using Remote Deposit Capture", February 2009, Celent Research, Boston, MA, USA

8   "The Branch Expansion/Remote Deposit Capture Paradox", in *Bank Systems and Technology Magazine*, 3 April 2009

9   "E-banking used by four in five online US households – survey", Finextra.com, 15 July 2009

10   "The decline and fall of Britain's high street", *UK Times*, 19 September 2008

11   "Aussie consumers choose Internet banking over ATM, phone and branch", The Nielsen Company, 26 April 2007

12   "ANZ losing branch staff as online banking takes over", *NZ Herald*, 25 September 2008

13   Case Study: Rabobank relies on NonStop systems for Internet banking, HP and Rabobank, 2006.

14   "Consumers seek more self-service options due to pressures of price and time", NCR 2007 & 2009 Self-Service Consumer Research

15   Take the virtual tour of the Q110 Branch at http://www.q110.de/tour/tour_ java.html?plan=1

16   Deutsche Bank Annual Report 2005

17   Jyske Bank – Jyske Differences branch concept. http://www.jyskebank.ek/jyske bankinfo/home/home/220771.asp

18   "Why Wells Fargo Bank is Different", by G. Pascal Zachary, *Business 2.0 Magazine*, 12 June 2006 (http://money.cnn.com/2006/06/09/magazines/ business2/bankdifferent/

19   Addison (http://www.addison.com/pages.cfm/what-we-do/ simplification/wells-fargo.html)

20   Wells Fargo Annual Report 2004, page 7

21   Rabobank.com, *ING Direct Annual Report 2008*

22   *South China Morning Post*, February 2007

23   ABN Amro Annual Report 2006, 2008

4 "Please Hang Up and Try Again"—Contact Centres and IVRs

The Rise of the Call Centre

UNFORTUNATELY, we sometimes just end up getting bad service from call centres. I don't know if you've ever sat in a call centre for a few weeks taking calls, but it can be hard, depressing and soul destroying work unless you have a really positive view on life and the world in general.

In the UK, surveys show that call centre staff turnover on average is around 25 per cent per annum but can peak out at between 60–80 per cent.[1] In locations such as India and China, staff turnover averages out around 50 per cent and can even top 100 per cent a year in companies that have poor HR policies and staff support mechanisms. Many contact centres employ young university students who are looking for some quick cash and a job where they can choose their time commitments. Such employees, however, will generally leave as soon as they finish their studies and move to their profession of choice. Staff turnover is just one of the issues facing contact centres in the shift to **BANK** 2.0.

The key issues affecting bank management regarding contact centres today are pretty uniform whichever geographical location you may be in. They include:

- **Capitalising on sales opportunities.** The number one tactical challenge bar none—ensuring that staff know why they are recommending a specific product or solution to a customer. For example, why are we offering this guy a Gold MasterCard when he already has a Platinum Visa?
- **Consistency and quality of communication.** The ability to deal with customer enquiries, contacts and sales across any

| Capitalising on sales opportunities | Consistency & quality of communication | Quick access to the right information | Improving IVR utilisation and effectiveness |

| Making new technologies pay | Green issues | Customer advocacy | Finding the balance in outsourcing | Staff turnover & employee engagement |

Figure 4.1 The key issues facing contact centres in the BANK 2.0 world

channel, regardless of which point you enter the conversation. Email handling and regulating customer communications are also critical to avoid unnecessary issues.

- **Quick access to the right information.** This starts with being able to respond quickly to the most common customer enquiries upfront (prioritise call centre "dashboard" navigation, based on top call types in the last six months for example), plus an overview of customer's bank-wide portfolio, next best sales lead, last contact with the customer, etc.
- **Improving IVR utilisation and effectiveness.** IVR (Interactive Voice Response) systems were first brought in to reduce the cost of, and the load on, our contact centres. Now they are seen as the bane of our customer's existence, but it is possible to improve the way the IVR system works. Included in this we'll look at speech recognition—exploiting the value of speech as a possible replacement for touch-tone IVR.
- **Making new technologies pay.** Text chat, video chat, Skype support sound great in practice, but making these viable, profitable and with engagement options is important.
- **Green issues.** Persuading customers on the phone to switch to e-statements, e-receipts, etc.

- **Customer advocacy.** How to generate higher levels of advocacy amongst callers. Advocates of call centres talk of genuinely helpful, personable, friendly customer service agents who can empathise and create rapport with them. Enough said.
- **Finding the balance in outsourcing.** The trade-off between offshore cost savings and regionalisation vs onshore (perceived) better quality.
- **Staff turnover and employee engagement.** Providing career/skills progression in a call centre and improving staff satisfaction.

It is possible that something is broken with our call centres. Staff turnover in the call centre is the highest of any department in the retail institution. Problem resolution is often not successful. Customers find the current IVR processes generally frustrating and difficult to navigate. The common customer perception is that calling the customer service centre may or may not actually get you a solution to your problem at the very best of times. There is simply no guarantee of service through this channel, and it is not as cost effective as it once was.

A number of issues have contributed to the deteriorating perception of contact centres. Take the example of an ongoing problem that you, as a customer, have tried to resolve but were unsuccessful. Generally, you had to repeat the entire case history or story to the new CSR (Customer Service Representative), regardless of how many people you've spoken to beforehand. It doesn't matter if you've gone into the branch and tried to have the issue resolved, or how many times you've rung the call centre, you always get stuck back at the beginning. It is a rare instance of service excellence where an individual within the organisation might actually take ownership of your problem and do the work required to get back to you with a solution. Very rare indeed.

In other instances, there will be a communication disconnect between the call centre and other channels, or even within departments that are supported by the call centre team. The credit card team isn't talking to the preferred banking team, so when the cards team makes the call to put a stop on a customer's credit card, inevitably the preferred banking call centre gets a call from an irate High-Net-Worth client. The preferred banking team has to start at the beginning because they have not been informed of the decision by the cards team, and when they finally get to the core issue, they have to refer the client to the credit department in any case. The customer is not a priority in this process. Otherwise, the preferred banking team would have been notified first, and given the opportunity to call the customer BEFORE the stop is put on the card.

Yet contact/call centres have been around in various guises for more than 20 years now, so why do we still have such issues? Why the disconnect with the customer? Let's see if we can tackle the issues above in some sort of structured manner. Why don't we start with generating revenue.

Capitalising on Sales Opportunities

Initially contact centres were designed to reduce the cost of servicing existing clients. Over time, however, management realised that since customers were already calling in and we knew a fair bit about them, we could leverage that opportunity for selling. Outbound call centres also started to spring up, and became very popular for telephone companies and others that were trying to reach customers who were responding less and less to traditional advertising and direct mail offers, and those who were hard to track down except on an evening when they were at home taking a respite from their hard day at work.

So popular was the phenomenon of outbound sales that we actually overdid it. In industry terms, we now refer to this outcome as "burning the list" or "reducing database effectiveness", but needless to say, we did so much outbound sales activity that in many countries we now have to regulate how and when organisations can call consumers because of widespread consternation over telemarketing. We even have do-not-call lists and mechanisms available for individuals to opt out so outbound sales centres are restricted from calling these individuals. On sites such as YouTube and Twitter, we hear these classic stories (sometimes with actual recordings) of individuals who have received the most persistent of sales calls, or customers trying to cancel their accounts or plans to little avail because some poor CSR has KPIs (Key Performance Indicators) that are built around keeping the customer at any cost.

So while outbound sales centres have been taking the brunt of customer dissatisfaction in the last few years, the reality is that they can still work with just a little bit of focus and thought.

Optimising sales performance in contact centres comes down to two key elements—better targeting and better sales conversations. When we started with telemarketing it was always simply a numbers game. Today we need to know who we are targeting, and we need to know what they need, so that we can fit the right product to the customer.

Better targeting

Firstly, the objective has to be about the right product at the right time to the right customer. This takes some intelligence, but increasingly we have great pools of information available about our customers; we just have to tap into the information and make use of it. Secondly, we need to be able to have a conversation with the customer that appears more like a *service* exchange than a sales exchange. How is this possible? Well, if the offer is presented as a solution to a potential problem for the client, then they equate this with better service from the institution, rather than a pure sales pitch. This is really the only way to tackle the psychological hurdle of the "sale". Move from selling to **service-selling.**

Any time I present a badly positioned or poorly selected product to a valuable customer, the more likely I am to fail, not just once, but in the future too. Yet, time and time again today, we see inbound call centre managers saddled with the "offer of the month" to position to customers who call in. As effectiveness has reduced, we increase the offer of the month to perhaps three or five different products that the CSR can choose from on the fly, depending on what he knows about the customer, but it is still hit and miss.

For **inbound solutions,** we need to have intelligence built into the system in respect of the next best offer or the next best action for the customer. This requires the implementation of customer analytics, business intelligence and offering management solutions to take a range of offers each month, and match the right offer with the right client. It also requires each of the product teams to come up with a range of offers each month that can be triggered by key data points for specific customers.

Most product teams are used to running just one acquisition-type campaign on a new product three or four times a year, and it takes a great deal of work with creative agencies and so forth to pull it off. Thus, approaching these teams and asking them to come up with core benefits or sales messages for 10–15 different offers each month requires a fundamental change of philosophy internally. In later chapters, we'll talk about the creation of a customer dynamics team that can assist here.

There are three elements that make an "offer" viable for use in the call centre. Firstly, an offer is not a campaign; it is a simple sales message tailored to a market segment or client profile. Secondly, it is designed to be actionable—with a simple, understandable proposition that has specific benefits for the client it is offered to. Lastly, it does not require a great deal of preparation or a steep product learning curve in order for a CSR to be able to pitch an offer. In fact, he or she can probably just read it off the on-screen script.

Figure 4.2 Essential components in successful telesales

Better sales conversations

There are three classes of offer types and scripts that need to be developed for the sales conversation regularly—acquisition, cross-sell and up-sell. These need to be reinforced with training.

Acquisition is for targeting new customers and is the most difficult to be driven from an analytics perspective, but there is a way. For example, we may have customers who have taken a credit card with us but have no savings or current or checking account. By looking at their card history, purchases and payment history, we can get a picture of potential needs. Maybe at certain times of the year they make certain big purchases that might be better funded by a lower-interest-rate credit facility, for instance.

For **cross-sell** and **up-sell,** we already have all the information we need in the bank's various systems, but we rarely use this information effectively. Primarily this comes back to the systems disconnect issue we will talk about later in the chapter, but the issue is very much about learning from the data we have on customers to ensure we pitch the right offer to the right client.

Figure 4.3 Establishing effective sales conversations requires developing sales scripts in key areas

The analytics should be looking for opportunities around the following categories of offer:

1. **Products that the customer has purchased before** but currently does not utilise, such as general insurance or a term deposit.

2. **Products that a customer could use** due to transaction history or linked product activity, but doesn't, for instance, a customer who has a car loan with the bank but not a motor vehicle insurance.

3. **Products that improve a customer's life** or aspirational products, such as upgrading a customer from Gold to Platinum status on their Visa card, or pushing them into the "preferred" bracket even though they don't meet the minimum balance requirement.

4. **Alternative products that give the customer a better deal** than the current solution. For example, for "revolvers" who maintain a high credit card balance, an offer to shift some of the balance to a line of credit facility at a lower interest rate.

5. **Bundled offerings** or products that go well combined with other products, such as mortgage insurance or contents insurance with a new mortgage, car insurance with a car loan, or a Platinum credit card with a Mutual Funds investment.

6. **Future point-of-impact offerings** that are time sensitive or might be linked to a future action, purchase or trigger. For example, a tax loan at tax time, or a great travel loan deal when the customer takes his annual family holidays.

As customers, we rarely receive such well thought out offers. I have a spread of products with various financial institutions that could easily be consolidated with one or two institutions, but the fact is that often I make the choice on a product because of expediency and because my primary institution relationship is not anticipating my needs.

For example, I go to the local car dealership to look at upgrading my family car. I typically opt for some sort of hire-to-purchase or leasing deal with the dealer, and I normally go with the recommended financing option proposed by the dealer because I can sign up there on the spot. The same goes for the vehicle insurance if it's possible and not much more expensive than my current deal. But if my bank said they'd look after the negotiation with the dealer and fix all the details, and they'd match the deal offered at the dealership by their competitors, I would be happy for them to get the business and look after all the fine print for me—because this would be a great service. This isn't rocket science, but the bank should be ready for such opportunities. How? **Better customer analytics.**

The offer management or generation process needs to be a **dedicated function** within the bank. The marriage of the product team and the contact centre sales team needs to occur through a function such as a customer propositions or **customer dynamics team,** but one with real clout. For example, if the product teams aren't supporting the process, then the contact centre is not obliged to promote their product of the month. The customer dynamics team, however, can assist the product team with the right data and analytics so the crafting of relevant offers is made easier.

The product team can even second members of their group to the customer dynamics team to ensure the offer messages and positioning are correct.

With the contact centre itself, the offers need to be delivered contextually on a single-screen view or dashboard where the CSR can see the most relevant deal or offer based on the customer's existing product footprint. At the end of the day, the CSR has to feel that he knows enough about the product to sell it. So, making the key benefits to the individual customer clear is essential.

Staff Turnover and Employee Engagement

While many calls to a bank's call centre are routine, such as enquiries on account balances or whether a cheque has cleared, the call centre team also gets to deal with the most frustrated of customers who happen to be at their wits' end. It is damn hard work. Sometimes, just sometimes, staff will let these frustrations show to customers.

A friend of mine recently related the story of an occasion where he wrote out a cheque for the discharging of a loan and gave it to the loan officer at his branch. The cheque got all the way up to the loan department and someone realised that the amount on the cheque was wrong due to a calculation error and it could not be processed. This was in the United Arab Emirates where the bank in question only had a few branches in the country, and for my friend this was a reason he generally didn't commute the hour or so to the branch on a regular basis, choosing instead to do the majority of his banking online or via phone and ATM.

The bank called and told my friend that he would have to go back to the branch. Actually, to be honest, he wasn't totally sure that this was all the bank's fault. To this day, he has a sneaking suspicion that maybe he wrote out the cheque incorrectly. However, he just wasn't able to spare the time during an extremely busy working week. The bank suggested some alternative methods, but all of these still required him to visit the branch. He asked the bank if they could instead do an electronic transfer. "No, we need to see your signature for that." "No, we must have a cheque, sir." He was pretty frustrated, and evidently so was the CSR. So he just told them if they really wanted the cheque so badly, they were just going to have to send a driver or courier to come pick it up for themselves. The CSR's response, and I quote this verbatim: "What kind of bullshit is this, sir?"

Sometimes, frankly, customers are just really difficult to deal with, and even the best of us might crack under such unrelenting and persistent pressure. But as a bank, we should realise that often it is the outmoded processes and compliance requirements that produce such responses from jaded clients like my friend in the above example. While the CSR's response was unforgivable, the fact is that the entire episode was probably completely avoidable.

The customer is simply thinking: hang on a second, this is supposed to be the 21st century, and you guys are asking me to come down to the branch and sign a physical piece of paper when you have had a relationship with me for the last six years and you have a million other ways to verify my identity and get this done! But the compliance rules of the bank dictate this is the box the frontline office must have ticked to proceed, so as a customer I have to go out of my way for your antiquated processes.

Many times these service issues that the bank creates are just that— service issues the bank creates.

Getting staff retention right

It is no wonder that in today's fast-paced, changing work environment, attrition rates are high. Nowhere is this more evident than in a customer contact centre. By its very nature, the atmosphere is hectic and intense. Phones ring constantly and emails abound as customers by the hundreds seek answers about the organisation's products or services. The environment is usually tightly structured with the primary focus on quantity and speed. In this highly charged setting, it's not surprising that employee retention is an ongoing challenge. According to Gartner Group, overall attrition for contact centres globally is averaging 15 per cent per month.

High turnover, however, has other negative effects beyond just the cost of replacing staff. Turnover creates morale problems in the contact centre, increasing the likelihood that other staff will be affected by the departure of friends and colleagues. It reduces staff productivity and lowers overall service levels. So the very thing we are seeking to achieve, namely good service response for customers, is negatively impacted by our biggest management challenge—staff turnover. Another report on four high-turnover industries by Sibson & Co. found contact centres experienced turnover rates of 31 per cent annually in the US. The same report showed that high turnover reduced call centre earnings by 43 per cent, leading to an estimated industry-wide cost of $5.4 billion. So turnover is expensive in more ways than just the cost of attracting, rehiring and replacing staff.

There are many potential causes for staff turnover. Certainly, economic conditions as well as factors such as market and competitive pressure affect turnover rates. These more general causes for *involuntary* turnover are not easy to manage because they often occur independently of the organisation. However, there are certain causes for *voluntary* turnover associated with contact centres, such as non-competitive compensation, high stress, unpleasant physical or interpersonal working conditions, monotony and poor direct supervision, that can be better managed.

One option for addressing the issue of high staff turnover in contact centres is to try to re-engineer the job to eliminate the negative characteristics, but given that most of the negatives are associated with the job itself, this may not be entirely feasible. We can also attempt to screen out potential "quitters" during the hiring process by asking the right questions. The simplest way to do this would be to use the information regarding negative job characteristics—those cited by staff who leave, for example—as part of the pre-employment screening process in order to identify job applicants who are likely to be affected by such negative issues.

We discussed in the introduction the fact that staff employed in a contact centre often see the job as a temporary fix rather than a career path. This is obviously one of the key issues to combat. If someone comes into a job expecting to leave it in the not too distant future, retention is always going to be an uphill battle.

According to research conducted by Thames Valley University (UK), at least 60,000 of the 171,000 BPO (Business Process Outsourcing) workforce in Indian outsourced call centres change jobs every year.[2] About 80 per cent of this workforce says they are looking for better leaders to work with. The career progression challenge means that team leaders often want to be upgraded to supervisors, quality professionals or operations heads, and aren't interested in "detail" day to day if it gets in the way of their career path. This and the lack of employee engagement are key issues that need to be tackled.

Employee engagement and creating a service culture

Team leaders need to be taught to lead more effectively, and they need to demonstrate vision. It's not enough to say "we are customer-focused" or "service is our business". The culture of service needs to start at the top and filter down through the organisation every day. If you create an organisation that lives and breathes a positive service culture, it will also create a positive working environment.

The undisputed leader in improvements in customer service globally is a friend of mine, Ron Kaufman. Ron has created what can only be described as a global movement in world-class customer service education. He runs a global education programme through his aptly named "UP Your Service! College", and he's a prolific author and speaker. Oh, and he's just an all-round nice guy, as you would expect.

For years, staff in contact centres have been saddled with the baggage of an institution that believes they are organisational refuse or hangers-on in the big world of retail banking. It is a job that carries little prestige. In fact, such staff might even be scorned by serious "bankers" as "those guys in the call centre". This is a tough monkey to shake off your back if you are sitting in the contact centre trying to motivate yourself to take that next call from an irate customer.

Figure 4.4 Ron (first row, second from left) with staff of a Middle-East based service organisation (Credit: UP Your Service! College)

The first thing contact centre and frontline staff need to succeed in doing is to take pride in their work and in their achievements, and build a desire to do great things for customers. Ron created the UP Your Service! College as a global mechanism to educate and motivate contact and frontline staff from around the world. It has been an outstanding success for organisations such as Microsoft, Wipro, Singapore Airlines, Dubai Holdings, ELQ (Holland), HP, Xerox, Temasek, FedEx, Dell, Raffles Hotel and the American Club. He cites a number of key success factors that are essential to creating a strong service culture and ensuring employee engagement:

- Common service language;
- Fundamental service principles (not just a catch phrase);
- Key learning points;
- Application to the job;
- Practical action steps; and
- Internal leaders "certified" for action.

The "UP Your Service" framework for service excellence is built on the concept that excellence is a moving target, and for an organisation to reach the target, you always have to be moving up the service excellence ladder.

The concept of "Moments of Truth" is one that is bandied around by consultants and experts alike as the foundation of service excellence. I don't want to steal too much of Ron's thunder because you really should read his books for yourself, but Ron has taken this concept and built a much more tangible set of motivational drivers for internal change. His moments of truth turn into "Perception Points™"—those opportunities for creating a service perception in the mind of our customers that occur at each and every service engagement.

His concept of service excellence is in the form of a six-stage ranking that tests whether your service response is in the range of Criminal or tops out at Unbelievable! (Figure 4.5) There aren't many banks today that could claim their service is even in the range of Expected on Ron's scale, let alone Unbelievable! The key here is engagement of every employee from the janitor to the CEO in the culture of service excellence.

Figure 4.5 Ron's framework for creating a superior service culture
(Credit: UP Your Service! College)

There's a nice anecdote about John F. Kennedy visiting NASA Headquarters during the 1960s as the US made its run-up to the Apollo landings. JFK stopped to talk to a man holding a mop in the halls of the NASA administration building. He asked the man what he was doing. The man, a janitor, replied, "I'm helping put a man on the moon, sir." Is your staff helping to put your customers over the moon in respect of service expectations? If not, you may need to up your service culture.

The fact is that every single employee working in the bank needs to have done some time in the contact centre to be able to empathise with both customers and call centre staff. It is a great way for cadets who are on a career track to cut their teeth in the organisation for the first year listening every day to customers and their needs, concerns and issues. You have to be ready to deal with any product enquiry within the bank and tackle day-to-day issues that really matter to customers. You also get to know more about the organisation as you are constantly firing off emails or transferring calls in an effort to solve customer problems.

By getting to know other contact centre staff as well, career track employees will not only have empathy for the contact centre team, but when an issue comes up to your desk from the contact centre floor, they'd be more motivated to try to assist in resolving it and passing the resolution back down. It helps institutionalise the service excellence culture even further.

The reality is that staff turnover is increasingly a fact of life, not just in contact centres, but in all sectors. Employees are generally staying with the job for shorter periods. By the time my kids are in their thirties, they will probably have worked for six different companies, staying on average for only 12–18 months at a time. So in addition to working to retain the best staff longer, in the **BANK** 2.0 paradigm we will have to get much better at increasing the productivity of new staff more quickly.

According to research by PeopleStreme, 77 per cent of employees are unhappy with their current job, and 39 per cent of employees who are not actively engaged suffer from work-related stress.[3] Might this have something to do with turnover rates in your organisation's contact centre? In contrast, an engaged employee feels more positive, believes in the organisation, understands and works with the organisation in respect of common goals, and takes pride in providing customers with outstanding perception points.

"Service is the currency that keeps the economy moving. I serve you in one way, you serve me in another. When one of us improves, things get better. When both of us improve, it really makes a difference. When everyone improves, the world grows closer together." Ron Kaufman, UP Your Service! College

Consistency and Quality of Communication

Banks are frequently talking about first call resolution strategies, call resolution quality, outbound sales strategies and so forth, but in actuality most retail institutions don't have a department-wide customer communication strategy or policy that ensures consistent service performance.

At any one point in time, a customer may have contact with multiple touchpoints, including the IVR, the inbound call centre, the outbound call centre, the branch, the ATM, the Internet, a relationship manager and a direct salesperson. Additionally, institutions are constantly sending information to customers such as statements (both paper and electronic), direct mail offers, transaction advice, PIN numbers, SMS and e-alerts. The problem organisations now face is how to regulate all this communication with customers across these disparate touchpoints while remaining cognizant of the needs and status of each individual customer.

The danger of unbridled email communications

Then we have email. While a cheap avenue of communication, email is probably the single most dangerous medium available for your staff to communicate with your customers. Why? Because probably the least skilled, least motivated and lowest paid member of staff who has a company email account is just one email away from creating a customer service problem of epic proportions.

In some instances, even full-blown marketing campaigns might go out to a whole distribution list with such eloquent prose as this classic example below inviting HNWI clients to an investment seminar in Hong Kong.

From: E-Marketing Team
To: Customer@asiaco.com
Subject: Email Invitation to Investment Seminar

Dear Valued Client and Investor,
In the success of the China's entry to the World Trade Organization (WTO), how would our Fund Manager looking in the world's leading coverage of Greater China … and more. Please come to our upcoming Investment Seminar.

Topic: "China Market Reform"
Speaker: xxx
Date & Time: Saturday, 29 September 2007, 2:30pm – 4:00pm
Venue: xxx, Causeway Bay, Hong Kong
(For further details, please feel free to call our Investment Hotline at 9999-9999)

Obviously I've removed some details here to protect the originator (and myself!) of this wonderfully crafted email, which came from a major North American bank with its Asian headquarters based in Hong Kong. It's a classic example of lack of coordinated control over the core communication channels. The fact is that this email, penned in traditional Chinese, was probably first sent out to a supervisor for sign off before it went out and was translated by an IT person or junior staff member of the marketing department (because they needed to send out the same to their English-speaking customers). But no one verified the English version because the Chinese version was fine and had already been cleared by the departmental head.

Worse than this is happening every day as a result of contact centre and frontline staff communicating with customers one-on-one. Direct communications from staff without any quality control mechanism or policy on how to handle customer issues may invariably lead to communications that could effectively destroy a customer relationship in seconds. Take a look at the example below.

From: xxx
To: Rebekah Keen
Date: 5 October 2008 — 5:48pm

Dear Rebekah,

All I get to hear from you is always complaints. There are certain procedures we need to abide by. I am not surprised to read this mail from you as it was very much expected.

The Credit Department first reviews the initial application which is already forwarded to them and then they come back to us seeking more information if required. And for your information, I have several other Accounts with a substantially bigger deposits and operations and they are quite content with the services provided.

However, if you still feel the need for an Account Manager change, do call me whenever and I shall look into having that done.

Relationship Manager, Priority Banking Team

An email such as this should never have been allowed to see the light of day, whether or not the relationship manager felt he was in the right or wrong. The issue is that this has almost certainly permanently damaged the brand of this institution in the mind of the customer. There should have been specific training on how to handle the issue, and most likely the response should not even have come from the relationship manager in question, or if it did, it, should simply have been an acknowledgement of the customer complaint and a suggested resolution.

With an estimated 165–190 billion corporate emails distributed daily,[4] the sheer volume of electronic correspondence literally guarantees every company will experience a serious incident of non-compliance resulting from their use of email—whether they know about it or not! According to Harris Interactive, less than half the email users in listed companies comply with corporate email policies; many have not even read the policy.

One hundred and sixty-five *billion* emails every day. That means the average worker will spend 10 years of his working life just dealing with email. It means your company pays US$25,000 every year alone for your senior executives to deal with emails, 70 per cent of which are spam. This works out to over US$130 billion a year in cost to the industry. Still, 25 per cent of your workers are more likely to actively prefer communication via email than face-to-face. The psychology of changing customer behaviour means that by 2020 most of your employees will probably prefer non-personal, electronic communication over face-to-face traditional mechanisms.

This is why email needs to be managed with great care. Up to 70 per cent of critical business information ends up being distributed via corporate email at some point in time. In 2004, a survey by the American Management Association and the ePolicy Institute found that 20 per cent of the responding companies had subpoenaed employee emails in the course of a lawsuit or regulatory investigation.[5] So what needs to be managed? Email use policy should include:

- internal corporate policies;
- distribution guidelines (when to cc someone or when to restrict distribution, etc);
- acceptable use;
- acceptable content;
- privacy guidelines (which information on a customer should/ shouldn't be included in an open email, etc); and
- business workflow for specific scenarios (for example, complaints handling).

Often, both internally and externally, email actually creates more problems than it solves. A seemingly innocuous statement can often be misconstrued and turn into a major issue because it is read incorrectly by the recipient.

Although email is often useful, its overuse is leading to a deterioration of communication skills. Each electronic transmission is a lost opportunity for improvement in interpersonal development. We use email to communicate with everyone. As long as its use does not hinder the spread of information, it is convenient. When one-line emails are traded back and forth, however, it really is quite pointless. This is exemplified when friends or colleagues trade numerous emails in a day, trying to arrange a time to meet, discussing a project deliverable or meeting agenda or aimless banter over other colleagues who fell asleep at the last staff meeting. Picking up the phone is often much simpler and faster. I have a colleague who, I swear, has not picked up the phone to call me in over three years. He's also the sort of guy who will simply stop emailing for weeks at a time if he thinks your last email was too gruff or stern. It's a very dysfunctional way to work.

Issues abound when we rely solely on email because we are not picking up the verbal intonations common in spoken language. It is harder to determine what the other person really wants. Checking email every couple of minutes when waiting for a reply is a waste of time. Once there is a response, we are often so eager to send something back that we fail to look at what we write. Poor grammar is common in email and successive mistakes weaken our writing skills. The emergence of SMS-type slang and abbreviations in email and IM chats is further evidence of the decline.

A few quick tips on email etiquette within the corporate workspace:
- Consider if there is a need to reply to any email that is "cc'd" to you.
- Don't use email as a substitute for regular face-to-face contact.
- Don't use email for supervisory functions, such as feedback on job performance and disciplinary matters, unless it is a formal conclusion to a meeting involving a member of staff and dictated by HR policy.

Many central banks and regulators around the world recommend, or even mandate, that retail banks have very strict communication policies regarding communications with shareholders and other stakeholders.

> "Companies should design a communications policy for promoting effective communication with shareholders and encouraging their participation at general meetings and disclose their policy or a summary of that policy. Any departure from this recommendation is required to be explained in the annual report."
>
> Recommendation 6.1 of the 2007 (Australian Stock Exchange) ASX Corporate Governance Principles and Recommendations

Isn't it ironic, though, that 99.5 per cent (or more) of your day-to-day communication is, in fact, not to shareholders or investors, but to customers and that these remain totally unregulated? How many banks have produced or enforce a general communications policy in respect of customer communication? Admittedly it is tough to do, and increasingly so with blogs, Facebook, YouTube and the like.

If you are unsure about whether or not your bank needs a better communications policy, think about this. Not only should you have an updated electronic communications policy, but you now also need a social networking use policy and, if you are like IBM, Sony and others, you might even consider a set of guidelines for your staff when they are operating in virtual space. IBM's Virtual World Guidelines include 11 separate recommendations on working within virtual space and representing the IBM brand responsibly through your avatar or digital persona.[6] We'll talk more about avatars in Chapter 10.

Many banks currently ban the use of sites such as Facebook, YouTube, even webmail such as Hotmail and Gmail from the company network. The corporate mantra is that employees might waste huge amounts of time, and could disclose company secrets through this medium if misused. However, with customers, partners, colleagues, friends and family using this medium, a blanket rule of not allowing use of these sites may not be the most prudent approach. After all, there is nothing to stop staff from using these sites after hours in any case. Additionally, research shows that organisations that allow these technologies actually have *higher* productivity than those that don't, and they have better engaged and happier staff.

Simply, the Chief Information Officer needs to implement a proper **Electronic Communications Policy** governing use of email, posts on blogs, use of social media, and participation in virtual worlds and other media. If you don't educate staff on using these new media properly, then you are leaving public forum interactions on your brand to chance. Dell has taken the opposite approach, and now actively engages customers via Twitter, for example, in customer service dialogue—with great success.

Quick Access to the Right Information

Systems disparity in today's call centres

This is both a design and organisational issue. In many instances, the call centre simply doesn't have the tools to assist you. Firstly, call centre staff are required to have an understanding of every product and service path, so they are spread pretty thin already. Often, they aren't told about changes being made in respect of customers, and so the first they learn of a potential issue is when a customer rings up to complain about that problem.

Secondly, and interestingly enough, the choice of critical IT systems to support customers within the bank is not generally made by the IT team. If you want to get a new administration system for supporting the credit card team, then it will generally come out of the budget for the credit card department or product team. Thus, while IT can make their recommendations, propose guidelines, and even suggest the platform or architecture requirements, it is still basically up to the head of cards to make the call on the final vendor selection.

Thus, you get every different department at some point commissioning or purchasing an IT solution that fits its departmental needs, but it is basically a completely separate system to those of the other departments. So there is a cards system, a separate mortgage system and a dedicated CRM (Customer Relationship Management) system, but that doesn't have any account information, just customer particulars. The core system of the bank has the most up-to-date account balance and transactions, but that system is on the mainframe with a DOS-based window for doing enquiries and lookups. Don't even get me started on the system that looks after credit assessments for clients applying for a loan, or the system that looks after general insurance products.

So here you are, sitting in the call centre, and if a customer rings up to get the balance on his credit card, these might be the steps for the CSR sitting at the call centre desk. I've represented this in a flow chart for ease of description (Figure 4.6).

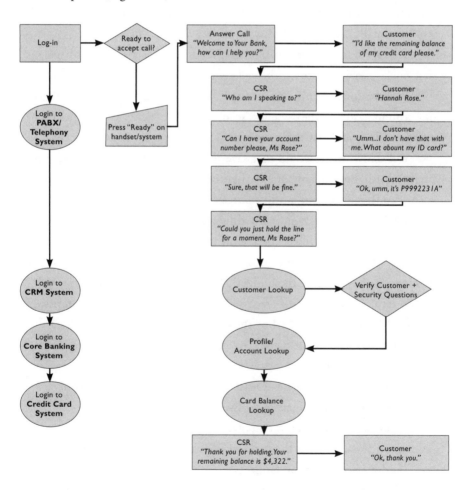

Figure 4.6 Call centre flow chart for simple credit card balance enquiry

Just for the simplest of enquiries, the call centre operator or CSR has to use at least four separate systems: the PABX/CTI Telephony Gateway, the CRM system, the Core Banking system interface, and the Credit Card system. Maybe you think that this is perfectly reasonable, but wait till you hear how many systems a customer has to use to find the same information on Internet banking? JUST ONE.

I log in to Internet banking, I provide my username and password (and probably a two-factor authentication requirement like a one-time password), and I'm in. On the first screen I see, I get all the balances to my account. Yet, within the bank, a call centre operator is forced to do all this work manually on a myriad of systems. Why?

When I was involved in the usability testing of a new set of screens in a single-user interface for the call centre of a global retail bank recently, its staff were amazed and astounded. "Oh, this is much, much better than our current system," they would exclaim. But the kicker is that these staff estimated it would cut their call times on average by 30–40 per cent on each call! How much would that save the bank? Millions of dollars every single month. So, can anyone here figure out why we still maintain systems with 13 different screens at call centres? Not me.

Why isn't anyone building **a single view** of the customer dashboard for call centres? Most likely the reason is that the bank is too fragmented and too "silo-ed" to organise such a complex project. Who would pay for it? How do they allocate the budget? Who would manage the project? It's all too hard, let's stick with what we've got. Here are some of the issues frequently raised or identified by the call centre team:

- Workflow processes broken by internal structure or lack of accountability.
- CSRs need to navigate at least 6 to 8 different screens most of the time to resolve customer issues.
- Call centre team is not empowered to handle applications properly, and often the handoff is a manual process that is hit and miss, resulting in failures.
- Percentage of first-call resolutions could be improved with the use of a tracking/contact management system.

Single screen customer dashboard requirement

With so many disparate systems and manual workflows, isn't it a wonder we don't have many more customer service disasters on our hands than we do? Today, a customer logging in to their Internet banking screen can get far better consolidated information than a CSR working in the contact centre. That is simply a poor business decision.

Before designing a single screen dashboard, it is helpful to classify the key components of information the CSR requires to handle the customer effectively. The components we have identified in our research are:

- customer related data and information;
- contact history (all channels);
- key frequent functions/transactions (assisted);
- key product applications/enquiries;
- account/relationship footprint; and
- sales opportunities.

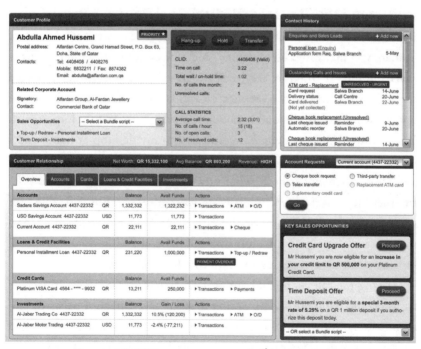

Figure 4.7 CSR contact centre dashboard concept
(Source: UserStrategy, Heath Wallace)

The call centre dashboard has to be integrated with the PABX and Telephony/CTI systems to improve both functionality and call resolution capability. The basis of the call centre dashboard is the call/customer relationship and tracking efficiency in resolving outstanding calls, and enabling key functionality to handle call resolution better, but also to make targeted cross-sell and up-sell recommendations.

In this way, all the key functions of the current call centre are available, along with a contextual understanding of call centre load, so that agents can respond in real-time to demands in an appropriate way. The integrated system can also log events for a CSR so that when a CSR takes a break or signs out of the system temporarily, we know what is happening. We can monitor how frequently a CSR is transferring calls to support staff, such as the credit department, so that we can assess whether the workflow on such tasks would be better handled by improving the information the CSR has through the dashboard.

Improving the "Invasive Virtual Resistance" System

Resistance is futile! When you call your bank today to try to speak to a human, you may often be frustrated. A classic example that I like to cite is that of my old ISP (Internet Service Provider) in Hong Kong, which provided me with broadband Internet service. In 2003, I tried unsuccessfully to reset my password for Internet access because I had inadvertently forgotten it. You see, my ADSL modem was connected to my desktop at home and I loaded the dial-up software on the PC, input the username and password and set it to remember both. That way, every time I needed to access the Internet I just hit the icon and bang—I was connected. But two years later when my PC hard disk went on the fritz, I could not for the life of me find that piece of paper with my username and password on it. So I thought I'll just call them up and get it reset. Not too difficult, right?

I rang the call centre and got an IVR system immediately. In fact, as far as I could tell there was no option to speak to a human. I persisted and eventually got down the list to option 6 or 7 where it asked if I needed technical support. Well, not really, but by the time I had listened to the IVR menu twice, I reckoned that was probably my best option. Within technical support were four or five options. The first was about setting up the Internet, the second about email support, and then there was the option for changing my password. It wasn't quite what I wanted, but it was close enough, so I hit "4" and went into the "change password" option. I then got a recorded message about how I could go to the company's website at www.ISP.com (not really their URL) and choose the "change password" option.

This is about where the problems start. First of all, I don't have Internet access because I forgot my password. Secondly, let's assume for now that I did have Internet access, it would probably require my original password to change to another password—so I was royally screwed. Out of desperation, I took my repaired PC over to my friend's place which had Internet access. I was able to fire up the Internet connection, get online and then bring up their website. I clicked on the link for "change password" and, as expected, I needed my original password. I looked around and eventually found the "forgot password" option. Aha, I thought, finally! I clicked on the hyperlink and guess what it told me: **"Please ring our call centre on 13…"**

At this point I lost it. The call centre was automated. I went through to an IVR system. It didn't have an option for resetting my password or speaking to someone. Believe me I checked every single path in that IVR menu structure, I hit "0" several times, I waited till the menu timed out, and everything else I could think of. The only option even close to what I needed was the "change password" prompt in technical support and that redirected me back to the website. Which directed me back to the IVR!

Eventually, I called the sales hotline. I got on to a person who wanted to know if I would like to subscribe to the company's new broadband service, and I explained as politely as I could that I was already a customer. I then explained my problem. The telesales officer was very polite and helpful and said, "I get these calls all the time." She then proceeded to transfer me to the technical support line (that I could not access from the IVR system) and they quickly resolved the problem by resetting my password.

So here I was—the customer—providing a workaround to a broken system, which the company must have known about. The system simply did not offer any solution to my difficulties if I followed the usual contact points or procedures. I probably could have gone down to their store to try and sort this out, but come on! This is an Internet service provider, after all.

The fact is, the IVR menu structure was patently designed without consultation of this organisation's customers. Secondly, I cannot imagine that my situation was particularly unique; in fact, the CSR told me it was not. Thirdly, what exactly was being achieved by not allowing me to resolve this solution? Would it save the ISP costs? No. Would it result in customer satisfaction? No. More likely, many customers simply would give up or even close their account in this situation. The potential for churn or loss of customers from this one process error was significant.

Someone wasn't doing their job. It could have been the call centre manager, or the IVR manager. It could have been the technical support team. It could have been any number of people within the organisation. But no one was looking holistically at the needs of the customer and working out how to address those. The fact that the website referred me to the IVR and then the IVR referred me back to the website, without my being any better off than when I started, was simply a disgrace. The fact that the telesales team was getting these calls ALL THE TIME (according to the CSR in the call centre) means that either no one was passing this information back up the chain, or if they were, it was being ignored. Obviously, no one was mystery shopping this sort of stuff either, otherwise surely someone would have addressed this, you would think.

These types of problems occur, unfortunately, because the IVR system today is seen primarily as a mechanism or symbol of cost efficiency within the institution, rather than a driver of improved service for customers. The bank's objective is to lower the cost of supporting you as a customer, thereby increasing customer profitability. So if you use the Internet instead of the branch I might save US$100–$500 on a typical cost of acquisition, for example. If I push you on to an IVR system, instead of letting you talk to a real person, again I have significant savings. If in the process customer satisfaction drops a couple of points, that is acceptable to shareholders who are looking at the next quarterly EPS (earnings per share) reports.

To make matters worse, in some places such as the US, France,

Australia and the UK, you can't even ring a call centre for free. They have 0800 and 1300 type numbers that actually charge you 10 times more than a normal local call for the privilege of calling.

Customers have already found workarounds to cope with the lack of human operators through IVR systems. Here are a few of the community efforts dedicated to bypassing IVR systems and getting to a real human:

- www.gethuman.com;
- www.phoneahuman.com;
- "How to hack a call centre telephone system" (YouTube video);
- "Deep Dial" to bypass IVR trees (SquawkBox/Alec Saunders);
- "Talk to a real person" cheatsheet (Techtarget.com); and
- many more.

Research has shown that an effective IVR system that enables customers to get a solution to their problem without having to speak to a CSR can save millions of dollars a year for the bank. For a call centre handling thousands of calls a day, the ability to shave time off the call response may result in savings of US$120,000 to US$1 million plus a year per second of average call time. Thus, the more I can get you to use the IVR, the more dramatic a cost benefit this technology will provide. Yet, there is much to suggest that customers are trying to find these workarounds to circumvent IVRs because they don't work the way they need them to.

So if IVR doesn't work today, is there a way we might be able to re-engineer the IVR menu to be more effective?

IVR menu design

When you ring your bank's call centre, you might be forgiven for thinking that you've stumbled upon an automated recording of its organisation structure—Press 1 for Retail Banking, Press 2 for Loans, Press 3 for Credit Cards, and so on.

Despite what you might think, IVR design is not exactly simple. Firstly, there are many menu options to classify and group into categories that customers will choose from. If customers don't understand the menu options presented, they are more likely to try to circumvent the IVR system.

Additionally, the more options you try to fit on the IVR, the more complex it becomes, so what we have to try to do is select the most beneficial options for both customers and the institution.

Statistically, around 70–80 per cent of the calls today to call centres of most banks around the world focus on just five to six key call requests. These generally concern current/savings account balance (about 60 per cent alone), credit card balance, recent transactions, bill payment and lost credit card. Additional popular transactions, depending on location and market, might also be special offers and promotions, activation of debit, credit or ATM card, and application status for a loan or credit card.

Companies primarily use IVR as a filtering mechanism to divert your call to the right agent who can help you with the problem in a most efficient manner, or in the best case scenario, solve your problem without human interaction. IVR best practice dictates that there should be no more than three layers between the customer and the CSR, but this has long been ignored by most banks. In their obsession to land the calls at the right agent's desk to improve efficiency, banks and other institutions have abused the IVR system. The IVR tree has now grown so many branches that if you find yourself on the wrong branch, it's often just easier to hang up and dial again instead of shouting for help.

Reviewing the following examples of typical IVR systems menu for various retail banks illustrates that the menu structures often bear absolutely no relationship to actual traffic or analytics. The menu design is taking place independent of call centre or IVR analytics, thus producing a menu structure that is not optimised based on actual likely utilisation. This is likely to add time to the call, frustrate customers unnecessarily and increase the rate of failures or dropouts—basically adding costs and reducing service perception for no good reason except that there is very little science applied to the menu design. Table 4.1 lists a few examples of IVR menus.

Citibank's IVR Menu tree (Figure 4.8, p.124) shows an admirable attempt to simplify the IVR flow based on likely transactions and your status as a customer. The IVR menu itself is limited to a reasonable list of options, never exceeding five options. Additionally, Citibank has instituted numerous individual dial-in options so that if you call the credit card line,

Table 4.1 Typical IVR systems across regions

Bank A: Asia	Bank B: Middle-East/North Africa	Bank C: USA	Bank D: Europe
Promotions	Account holder services	Products and services	Subscriber services
Rates enquiry	Credit card services	Online banking	General information
Products and services	Report lost/stolen credit card	Bill payments	Sales Agents
Phone, Internet or self-service banking	Promotional messages	Credit cards	Credit, debit and ATM cards
Branch or ATM locations	Loans	General information	e-forex customers
Others	Rates/product information	Foreign exchange	
	Credit card rewards	Others	
	End call		

you enter the IVR tree at the top of the credit card menu, avoiding the parent options of the common IVR.

There are some other design considerations, however. Take the example of your ATM card. If you look at the back of the card, there is a customer help line or support line. Ideally, this is the number you would call if you stuck your card in an ATM machine and it refused to give you any cash. In this instance, the IVR menu tree that this number reaches would reasonably need to offer an option at the top level that has something to do with getting help for a malfunctioning card or finding a suitable ATM. Broadly the suggested action plan for the banks in respect of IVR trees is:

1. Reduce IVR navigation to not more than four branches and four options per branch, with one of the primary options being the ability to talk to a CSR. If you have more than this, then the IVR is simply too complex.

2. Consider having different phone numbers for different IVR trees. For example, when you call the number on your ATM card it goes to an IVR tree that deals with ATM related enquiries ONLY.

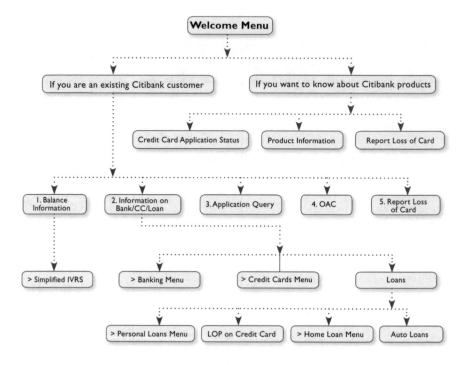

Figure 4.8 Citibank IVR tree (Credit: Citibank Website)

3. Inform customers upfront of expected waiting times for both self-service and service via live agent, therefore managing expectations.
4. Look for ways to flatten the IVR, like training agents or improving call centre dashboards, to handle a broader range of service calls.

Banks and other institutions have implemented the right tools the wrong way, as a result of which their applications help them keep track of data, increase visibility and improve efficiency on customer cost management. In the process, however, many banks have missed out on the relationship component as cost savings, and silos become the primary political consideration in the IVR deployment. At least for now, the conclusion is that good CRM and customer-centric IVR menu design are still a differentiation factor.

IVR prototyping and usability testing

Challenging conventional IVR design is a dangerous business. Call centre managers are in a very difficult position politically. Firstly, every division or business unit wants to ensure *they* have adequate representation on the IVR tree. Secondly, what the customer actually needs is a factor in design, but there are limitations on what can be put on the IVR menu. Lastly, the objective of the IVR is to attempt to solve the key needs of the customer first so that they don't have to talk to a CSR; but if they do, that they get far enough along the tree that we know who they are and what they need, and we can process their call quickly and efficiently when they get through to a CSR.

In redesigning the IVR, existing stakeholders who are invested in the current design will likely resist what might be seen as wholesale changes to the menu, pushing or opting instead for incremental additions to the tree. This approach produces a more complex tree as the menu is never restructured, just added to, which increases complexity over time.

Usability testing a lo-fi or low-fidelity prototype is an inexpensive way to validate a new IVR menu approach. This features the use of simple materials and equipment in order to create a paper-based simulation of an interface or system. Interface elements such as menus and voice prompts can be sketched on paper or created in advance using card, acetate or pens. To check the efficiency of the different models, most banks will probably want three separate prototypes to be developed and tested:

1. **Champion**—current IVR menu tree.
2. **Incremental Change**—business unit or product based.
3. **Prioritised by Customer**—call type and frequency from the call centre and Web analytics research.

The objectives should be to determine required functionality, to organise and prioritise the content, to determine the "call-to-action" components, and to ensure that the customer's journey through the IVRS is as smooth as possible. The two most important parts of the test itself are realistic scenarios (situations of use combined with user tasks) and representative users (test participants).

Usability tests should aim to have the participation of seven to eight customers selected from the target group with a wide range of demographics in order to assess the usability of the IVR system. The system is "simulated" by the facilitator reading out voice-prompts to simulate the IVR options, responding to user actions and observing users closely. Each user is given multiple tasks, one task at a time, to simulate their interaction with the new IVR structure. Users are asked to "vocalise" their response, that is, in response to "Press 1 for Account Balance, Press 2 for Credit Card Balance, Press 3 for Latest Transactions", they would answer "1" or "1 for Account Balance". The facilitator can ask participants to use their own mobile phone as a prop in the exercise. Video and audio recording can add further rigour to the test for review later by the various stakeholders.

This is a simple but accurate method of testing IVR system responses as it allows the test to simulate real time responses to audio queues. While it does not simulate the timing of prompts and the timing of responses as accurately as a full-fledged prototype or IVR system, as a comparative test of the champion/challenger approach, it will let the organisation know which is the most efficient approach to the IVR design.

Once again, customer feedback is the essential ingredient here. Ultimately, although the menu structures are different, customers will be able to get to the same information just through different information architecture (menu structure). The comparison of the low-fi prototypes will simply give you qualitative data to choose one approach over another.

Making New Technologies Pay

The last 10 to 15 years have seen a plethora of new technologies arrive on the scene that can potentially be utilised by a contact centre in providing support to customers. These include instant messaging via MSN, Yahoo and QQ, or Voice-over-IP technologies such as Skype, Google Talk and others. We've had webcams since the mid-90s, but these devices are now capable of some serious picture quality at a very low price. Additionally, we now have social media tools such as Twitter, Facebook and YouTube.

The issue for the institution is of course how to make such technologies cost effective. Would introducing these new technologies actually foster improved customer interaction? Would customers feel more comfortable communicating with the bank utilising these tools, or would they just stick with the traditional methods or channels that they've always used?

In many cases, these technologies have been around for years and not had a significant impact on our organisations. Many organisations do not even allow such new tools to be used within them. I know of at least five major international banks whose policy it is to deny staff access to webmail sites (such as Hotmail and Gmail), Facebook, YouTube and instant messaging tools. The argument is that if they allow staff access to these technologies, they are likely to waste valuable company time, or worse, let various trade secrets out. It could be said that if the organisation doesn't trust its staff to use such technologies, how can customers trust an organisation that doesn't even trust its own staff?

Let's take a quick look at these technologies, and assess the pros and cons for a service business such as a bank, particularly with a view as to whether they can be cost effective.

Skype, Web chat, video chat and instant messaging

IP-based communication tools are nothing new. At the dawn of the commercial Internet in 1995, Itelco released a piece of software called WebPhone®, and White Pine Software released CuSeeMe®. Neither worked very well and unless you had a flawless 56K modem connection, the likelihood was that your call would fail about 50 per cent of the time, or quality would be an issue. Sometimes you couldn't even get a connection. Compuserve, which became part of the AOL conglomerate, was successful early on with Web-based chat rooms and simple Web catalogue services. As the successor to what we used to call "bulletin boards" in the 1980s, these chat rooms soon got the reputation of either being an online meeting place for geeks and crackers (as in hackers), or a seedy online version of a strip club. Not very auspicious beginnings.

Things started to change in 1996 with the emergence of ICQ or I-Seek-U, a chat "client" from Mirabilis (now owned by AOL) that revolutionised PC-to-PC chat communications. PowWow had been released a few months before ICQ, but it did not gain the early support that ICQ did. ICQ enabled people from around the world to keep in touch with friends, family, co-workers, virtually free of charge. While a little limited in its capability, the success of ICQ was in the detachment of the chat "client" from the browser and the stability of the platform even over the most dubious dial-up Internet connection.

ICQ took off quickly. Within just nine months of operation, ICQ had amassed seven million users. In 1998, as ICQ surpassed 12 million users, AOL purchased ICQ for approximately US$400 million. ICQ went on to surpass 100 million users by 2001. AOL already had their own client called AIM or AOL Instant Messaging, but it had not taken off like ICQ despite its history being enshrined in the very early days of the Internet through Commodore 64's Quantum Link service (which became America Online). Excite, Ubique, IBM Lotus and Yahoo quickly followed suit with their own "messaging" clients, and eventually Microsoft got its act into gear with the Microsoft Network Messaging Client, now simply known as MSN or Live Messenger.

In the initial variations on the ICQ theme, the method and technologies were pretty much the same. All these applications used a form of IP-based protocol to enable connectivity; they didn't need to run within the browser (so no problems with clunky HTML and Java functionality), and they worked on a separate port to the browser and email, ensuring that you could use your messaging client at any time and instantly. Hence, the phrase "instant messaging".

As IM technology grew in popularity, the promise of these clients morphing into virtual telephone equivalents was bandied around. But in 1997 and 1998 as the dot.com bubble started to ramp up, VoIP (Voice over Internet Protocol) was still largely a concept. Telecommunication companies were violently opposed to the idea of VoIP, or making free phone calls over the Internet, and would relish the successive failures of these IM clients taking on the challenge of phone calls online.

Early VoIP attempts were actually started back in 1995 with VocalTec releasing a very early version of VoIP in a product they called Internet Phone or iPhone for short (not *that* iPhone). Later, TDSoft launched a VoIP gateway or softswitch technology. However, it wasn't really until

2004 that VoIP was integrated with PSTN (Public Switched Telephone Network) enabling calls from computer to a normal telephone, and even then it was not 100 per cent reliable.

Skype emerged on the scene in August 2003. Skype was written by Estonia-based developers Ahti Heinla, Priit Kasesalu and Jaan Tallinn, who had also originally developed Kazaa, a very popular peer-to-peer client similar to Bittorrent. Skype was a technology that from day one pretty much delivered on what it promised.

In economies where monopolistic telecom providers existed, it was not uncommon to hear of the local regulator banning VoIP. In September 2005, SkypeOut was banned in south China, with regulatory issues cited and these services remained blocked well into 2008. China Telecom, the main phone line operator in China, simply described Skype's online services as "illegal". Mexico's Telco incumbent, Telmex, blocked Skype and Vonage, and Deutsche Telekom did what it could to make life difficult for the VoIP services too.

Today in the United Arab Emirates, the Skype website is still blocked on the basis of "moral and religious" grounds. For years, the local operator Etisalat invested heavily in proxy technologies that would not only block the Skype website, but also attempted to block the protocol used by Skype (largely unsuccessfully). In May of 2008, the new competitor for Etisalat, Du, also blocked SkypeIn/SkypeOut services and the Skype.com website despite the fact that they had been allowing access to these services for approximately 10 months from the time Du commenced operating in the UAE. It was obvious that this was simply an attempt to force the 90 per cent of residents in the Dubai emirate who are foreign nationals to use primarily the operator's networks for overseas calls, instead of being able to use free PC-to-PC or PC-to-Phone VoIP services.

Efonica, a VoIP company operating their business out of the UAE, has their site blocked within the UAE! The 100-million-subscriber Vonage, as well as Net2phone, Webphone, DialPad, Babble, Go2Call, GizmoProject, IConnectHere, Lingo, MutualPhone, Netzero, Nikotel, Packet8, QuantumVoice, TeleSip, TerraCall, and doubtless, many others, also have their websites blocked in the UAE on "moral and religious" grounds.

Despite Etisalat and Du's best efforts, blogs and forums around the UAE abounded with instructions of workarounds, VPNs, cracks, onion routers, IP-blockers and other vehicles which would allow residents to bypass the telco's restrictions and allow VoIP protocols and clients to work. The Telecommunications Regulatory Authority in the UAE has declined to comment on why this is the case, only citing earlier instructions when VoIP was restricted by law. The website for Skype is still banned or blocked by both telcos on the basis of religious and moral grounds, although residents of Dubai simply download the software via other download sites via P2P, or get relatives to email them the latest version.

Commenting on this trend of blocking online websites, ECIPE research has come to the conclusion that countries subscribed to the WTO may, in fact, be in legal breach of WTO GATS provisions. Let's see how the UAE, Chinese and Saudi governments respond to such pressures from the WTO in the future:

> "... There is a good chance that a [WTO] panel might rule that permanent blocks on search engines, photo-sharing applications and other services [in respect of Internet censorship] are inconsistent with the General Agreement on Trade in Services (GATS) provisions, even given morals and security exceptions."
>
> European Centre for International Political Economy[7]

The global success of Skype was really a lesson in simplicity and was a lesson for telecom operators: that because you own the network doesn't mean you have an unasssailable monopoly that protects you. Technology innovation can remove barriers in an instant.

Skype was super easy to install, it was FREE, you didn't need any special equipment or plug-ins, it was very easy to operate and configure, and it worked every time. Skype took the Google approach to their client design. Usability on the software was fantastic and it just did one thing really, really well—which was VoIP. Skype took off like wildfire. Today, it has over 20 million *concurrent* users. It was acquired by eBay in October 2005 in a deal worth US$2.5 billion. In later versions, Skype allowed users to make calls to existing PSTN and Cellular telephones (SkypeOut), and to create a number in various locations where a call to a SkypeIn telephone number would be redirected through to the Skype client. As a result, Skype quickly proved to be a huge success. Yahoo, MSN and others have tried to emulate Skype's success, but it really was the killer app for voice communication over the Web. The only potential competitors are MSN and Google Talk, but even Google Talk's hype was short-lived.

In China, the Chinese government placed restrictions on MSN, Yahoo and other IM clients early on, suggesting they may be a risk to state security. An entrepreneurial group in China called Tencent produced what was marketed as a "China-approved" messaging client called QQ, launched in February 1999. It grew rapidly in local popularity and Tencent Holdings was able to support an IPO on the Hong Kong Stock Exchange in 2004 as a result of its tremendous growth. Due to its impressive Chinese user base of more than 440 million active users, QQ is now the most popular IM client on the planet. While the attempts on restricting other clients were abandoned, QQ had already become the dominant client in China due to its local language support and local buzz. In 2006, when the Hengchun earthquake severed trans-Pacific underwater cables, QQ gained further support when other IM clients suffered failure because of the lack of local server support on the mainland.

QQ is really the only case of a non-mainstream client succeeding in a specific geography. Outside of China, QQ's usage is extremely limited, with some limited exposure in South Africa. Yahoo retained a great deal of popularity across Asia also, and in Japan, MSN and Yahoo really were tied for popularity through the first half of the decade. But in Europe, the US, UK and most other geographies, Skype and MSN still dominate. ICQ had largely gone the way of the dodo. Why ICQ failed to dominate when it had such a propitious start, and why Skype still dominates to this day is a discussion for another time. Let's just say ICQ was probably underleveraged and under-marketed, while Skype's timing was perfect and their viral distribution strategy flawless. As ICQ was absorbed by AOL, which had their own AIM client, there were issues over which client should receive support from AOL as the preferred interface.

As of October 2009, QQ topped out with 990 million registered users and 440 million active users. MSN (or Windows Live Messenger) comes in second with 330 million users (June 2009); Skype comes in at third place with 309 million users (April 2008); and Yahoo comes in at fourth place with 250 million active registered users (January 2008).[8] AIM is popular in the US, but struggling globally with less than half Yahoo's penetration—around 120 million users.

My dad describes using MSN or Skype as "talking on the computer". My kids call it "talking on the Internet". But whatever you call it, the sheer volume of user take-up of these technologies means that banks, as a service provider, better start thinking about utilising these technologies because they are much cheaper than traditional methods and already widely accepted. Enterprise Instant Messaging and Enterprise VoIP are just something that you need to have.

Making it pay

For banks, such technologies represent an opportunity, but integrating this into the call centre may not be so straightforward today. Skype is working on such technologies and plug-ins for the corporate space, but the Web chat examples out there tend to be all unique implementations and can be a bit clunky. I suspect perceived reliability is the primary reason IM and Skype have not been better integrated into the enterprise contact centre landscape.

Why not just have a Skype link on the bank's homepage so customers can click straight through to customer service? Why not allow users to input their MSN or Yahoo Id in Internet banking to have the CSR respond to their query?

Let's deal with Skype and Google Talk first. Initially, you need to have the Skype calls handed off to the PABX or IVR, just like a normal incoming line. Currently, the telcos and the PABX providers don't really want you integrating Skype into their systems because Skype is a competitor to their PSTN and their own VoIP solutions. So, they are not going to make it easy for you. Secondly, call quality is not yet a guarantee; it would be like going back to the days of the analogue cellular phone. Lastly, it is likely that customers utilising this feature would use this capability in tandem with their Internet banking usage habits. The peak of Internet banking usage is typically first thing in the morning, and then late in the evening after dinner or a favourite TV show. In fact, in many markets, Internet banking often peaks around 10–11pm.

Putting all this aside, there are numerous solutions out there which integrate Skype into your enterprise PABX. Solutions such as Vosky, Skip2PBX, SkyPBX, SkyLYNX, PrettyMay, EZSky, S800 (Digital Switch), Zipcom and others have actually been around since 2007 and work just fine. Handling 20 to 50 simultaneous Skype calls is not a challenge for these technologies.

There are considerable benefits for going with this approach. Firstly, the call is basically free. Sure there is an integration and platform cost, plus bandwidth, but the incremental cost compared with existing PSTN infrastructure will still be much lower. Secondly, you have less contact failures and less customer frustration. For example, how many customers have gone to your site and couldn't find the call centre number, and had to ring directory assistance? By allowing customers to click through directly from your website, which fires up an instance of Skype and puts them through to the call centre, the chance of the customer perceiving your service as more efficient goes through the roof.

IM clients such as MSN, Yahoo and QQ still don't integrate particularly well with the call centre. There is no rotary call allocation system where customers can select the HSBC contact centre, and get put through to an appropriate CSR. If the bank integrated Web chat into the site, then it would be possible to manage this effectively, but Web chat to many customers just seems a bit primitive. If you gave them the choice to use their own IM client, they would be more likely to use something they are already using every day. Solutions, however, are starting to emerge for enterprise IM support. Vayusphere, Empatel, Talisma, and various IM gateways and SDKs basically mean you can integrate IM directly into the CSR's desktop.

Within the contact centre, the CSR may be having a hard time juggling an IM client and the required systems for handling the customer's call. Thus, screen design for integrating the IM would be required to make this work. This is technically feasible for most of the instant messaging standards and platforms. But given the disparity in systems we discussed earlier, without a single-screen dashboard solution, or giving the CSR a second monitor, this would be very clunky.

Crack these technology integration issues and you basically get a very cheap medium with which to provide customer support. As more and more customers move to transacting over the Internet and mobile devices, you reduce customer frustration and workload, and you improve customer service perception. Additionally, we can configure the VoIP or IM client to send the contact centre some tag (embedded in the on-screen banner button) which identifies the screen the customer is using. This might assist with faster call resolution also, further improving the case for utilisation of these technologies. It will only take one or two mainstream banks doing this and saving money before everyone else jumps in. Why not be the first?

Advances in speech recognition

Speech recognition technology has come a long way in the last few years. I remember using an early version of a system called Dragon Dictate. It was clunky at best, and any background noise such as a ringing phone or a barking dog could break the system.

Today though, we are seeing more and more banks integrating voice recognition into their IVR systems—Citibank, Wells Fargo and HSBC, just to name a few. But let's get this straight from the outset. Voice recognition rarely reduces cost on its own; it normally represents a nominal increase in cost. However, by the time we get this technology right, will we actually get cost improvements? In directory service utilisation and many booking systems, voice recognition has been shown to show reductions in total call times. In bank IVR systems, the gains are less certain.

IVR is a long-term viable solution for best practice, however. When customers can say what they need—for example, "help me with a payment" or "I've lost my credit card"—and we can respond appropriately, then not only will customer service perception rise positively, but our costs will be reduced as calls get diverted quickly to exactly where they need to go. The greatest criticism of IVR systems is that they take away the human element. When the bank IVR system responds in a more human fashion, then this perception will be reversed. This is undoubtedly where speech recognition is going over the next five to ten years, so the sooner banks get in on the action, learn the nuances of these systems and transition customers to the new approach, the better.

A recent development in speech recognition, though, can immediately improve customer service levels. Imagine you are a customer with a complaint or an issue. You ring your bank only to get the dreaded IVR system. After having to navigate 16 levels of the IVR, you are hardly going to be in a better frame of mind to speak to a CSR and hear their possible solution. Yet this is what happens every day in most banks.

New technologies in voice recognition enable us to determine if a customer is angry or unhappy. Remember, tone of voice represents a very large part of our verbal communication capability. Thus, emotive voice recognition allows us to flag an unhappy customer and immediately transfer him to someone who is trained specifically to deal with such customers. It does this through the combination of four different types of Acoustic/ Prosodic technologies, namely from Automatic Speech Recognition (ASR), Natural Language Understanding (NLU), Dialogue Manager (DM) and Context features.

Just a tip with this technology though. After the system detects the customer is upset, it would probably be better for the IVR not to say, "The system has detected that you are upset; we are transferring you to a highly trained specialist who is used to dealing with customers like you ..."

Going Green

Corporate social responsibility, climate change response and green-friendly corporate initiatives are all the go today. Incumbent upon us is the responsibility, if not the duty, to encourage our customers to be greener too. So we are spending lots on educating our customers about the benefits of e-statements and so forth. But are we ourselves going green and showing customers really that this is the way to work?

E-statements is a given for any bank's green strategy. Firstly, sending out statements by email instead of snail mail is an obvious cost saving. Secondly, customers can get their statements at anytime and the possibility of a statement going missing in the "post" is no longer an issue—as long as the junk mail filter is suitably trained.

Yet, going green is an oxymoron for most banks. While we are going green with e-statements, we are still relying on direct mail offers. While we are encouraging customers to go online, we reinforce a strategy where a physical application form is not only preferred, but in most instances, necessary for a new account opening or product acquisition.

Despite the fact that as a bank, I might have known you as a customer for 10 years and your salary comes into your personal account every month like clockwork, if you are applying for a new credit card or a personal loan, I am still going to ask you to print out and sign a new application form and give me three months of bank statements or salary history.

Legal and compliance are our biggest enemies in the battle to go green. The bank's compliance department is interested primarily in collecting a paper trail that reduces the risk of regulatory or legal repercussions in the event that a problem occurs. While this is a healthy corporate risk mitigation strategy, it is not an approach that either enamours the customer to the bank, or one that is compatible with a greener bank.

Tackling this issue of compliance department incompatibility with a green strategy is a big one. This has to start with a better informed regulator who promotes digital correspondence and records. The legal system, however, is built on the need for a paper trail. That said, digital signatures are legally recognised authority in most jurisdictions, and technology audit trails make tracking down the source of a document or file fairly straightforward. So while it is easier for the most traditional of organisations to stick to a paper-based compliance process, the reality is we can still generally provide a case history electronically that meets the regulator's needs.

If you are a bank CEO, task your compliance department with tackling this issue as a priority. Going green is going to save you millions and reduce process load at the front line by magnitudes. If you are a compliance officer, start learning to say yes or "let me work out something" instead of "no, we can't do that" a lot more often.

Better application processing

Picture this. I'm a frontline relationship officer and you are a customer walking into a branch to talk to me about investments, a personal loan for the renovation of your family home, or a new mortgage. Let's assume at some point I actually get to the stage of signing you up for a product or commencing the formal application for the product. What do I do? I reach over behind my desk, extract the correct form and start getting you to fill it out. If I am particularly conscientious frontline officer, I might fill out the form for you, just checking the details, and then ask you to sign.

For existing customers though, I am probably asking you for details that we already have on file and that you've already given the bank five or six times in the past couple of years. But this is inefficient, frustrates the customer, and is prone to manual entry error every time we rely on a paper-based form. If we must process the application on paper because our compliance processes are still stuck in the last century, then wouldn't it be more efficient simply to confirm the current address with the customer and then print the application form from the branch system so that all the details are already filled out and correct?

I'd like to complicate things a bit here though. Why is it that as an existing customer, when you log in to Internet banking, many banks now allow you to apply for a credit card, time deposit, mutual fund or a general insurance product without having to fill in a paper application form, but this is still a requirement at the branch?

So from a compliance perspective, there is already a precedent which allows you to apply for a product without the use of a paper form, but the branch processes are not updated to take advantage of this cost-saving strategy.

Outsourcing—the Case For and Against

Call centres gained popularity in the late 1980s, but by the 90s they were no longer a differentiator but a standard for most large service providers. The costs of call centres, however, were not insignificant and so many businesses sought to reduce the costs of these facilities. IVR systems were the first step, but as most IVR systems and menu designs were poorly thought out in the early stage, we succeeded in institutionalising the customer "workaround". Most customers wanted the option to speak to a customer service representative if they had the choice.

As a result, various offshore BPO (Business Process Outsourcing) and offshore contact centre facilities started to gain popularity. In the UK, US, Australia and other English-speaking countries, entire call centres were shifted to India or the Philippines to cope with growing call centre traffic. These industries represent major sources of income for India and the Philippines today. India remains the undisputed leader as a location for the outsourcing of call centres for the time being.

Service industries contribute about 52 per cent of India's GDP growth. Currently, India cannot cope with the demand for IT and BPO services with reportedly over five million jobs still to be filled. However, the global financial crisis led by the US recession is likely to have a significant impact. With 75 per cent of revenue for BPO in India coming from the US market, and an appreciating rupee against the USD, it is likely that operating margins of smaller BPOs will be squeezed hard. With BPO and IT outsourcing making up such a large part of the Indian economy, it is impossible to see a collapse of these industries anytime in the future.

Competition is now coming from a range of countries trying to get in on the action, namely Russia, China, South Africa, New Zealand, Mauritius, Fiji, Malaysia and Ghana. India has focused on BPO and IT, whereas the Philippines focuses more on voice services. Malaysia has zeroed in on transaction processing, and South Africa on non-IT BPO operations. Meanwhile, in the US, UK and Australia, telcos, banks and other service providers are facing a consumer backlash against the practice of outsourcing. Does it make sense for banks to outsource in this environment?

Pros—the case FOR outsourcing

The most obvious advantage for outsourcing is the massive cost reduction. An average call handled through an Indian or Philippine call centre costs about $0.65 per transaction (inclusive of labour costs), whereas that cost in the US is typically $2.28 per transaction or higher. When you focus on acquisition costs, the cost per acquisition can be much higher, depending on the product and ranging from US$45–$280 per customer onshore, and a fraction of that offshore. Outsourcing represents a much cheaper option to call centre response than a localised facility in a developed economy.

Training of call centre operators in these geographies is very efficient. Competency of staff, and therefore the ability to respond and solve a customer problem, is typically of a very high quality. Staff turnover is generally lower than in developed economies as staff are more motivated to keep their job because of local conditions. There are rarely issues of performance related to lack of capability or enthusiasm.

Capacity variations are much easier to manage in outsourced call centres. Typically based on some sort of SLA (Service Level Agreement), there are options for ramping up call centres as required from time to time. As most call centres operate for multiple clients at any one time, capacity is typically available on demand, whereas a local "in-source" option may have physical restrictions in respect of rapid growth such as telephony platform, lack of physical space, difficulty in hiring temporary staff and getting them competent in time.

As these offshore contact centres only focus on the one activity, this usually means they have the latest PABX and CTI technologies, great service capabilities and rapid response to technical issues, all of which may not be as efficient in a locally sourced option.

Cons—the case AGAINST outsourcing

The most obvious issue against outsourcing is that customers often say that they don't feel they are getting the same quality of responses to their issues as they had with local operations. While quality and technical competency of staff is very high in outsourced operations, customers often cite language difficulties and local nuances that hamper call resolution. Outsourced call centres have attempted to combat this by training call centre staff to sound more "local"—even running training classes to help CSRs adopt American or Australian accents. An analysis of 50 offshore and onshore call centres by the management consulting firm Compass found that problems in comprehension occurred 18 per cent of the time with offshore call centres, compared with 4 per cent of calls in onshore call centres.[9] These language and comprehension issues could lead to an extended duration of the call of between 39–105 per cent.[9]

The same report by Compass found that with wage gains in offshore centres of up to 15 per cent per annum, the economic benefits of outsourcing offshore are being quickly eroded. PowerGen and Lloyds TSB found the same concerns, and in 2007 moved their call centres back to the UK to improve customer service perception.

The biggest challenge yet, however, may come in respect of fraud prevention and identity theft. In October 2006 on a UK-based current affairs programme *Dispatches*, a Channel 4 senior reporter, Sue Turton, was able to buy bank account details, credit card numbers, signature numbers, passport and other sensitive information easily from middlemen in India working with offshore call centres. However, such fraud is possible even in the UK and the US because of the practice of appointing distributors and resellers for mobile phone sales, for example. Fraud remains a problem in the use of offshore call centres due to lack of regulation and controls, particularly at the lower end of the market.

Outsourced call centres also face challenges in adapting quickly to new campaigns or product releases. Often a product team will release a new product or new marketing campaign, but the word doesn't always get through to the call centre effectively. Thus customers enquiring on a new product or offer are not always successful in getting anywhere, which means that the campaign effectiveness is greatly reduced.

The alternative—homesourcing

In the US and UK, there are about 30 major companies including the likes of JetBlue, OfficeDepot, LiveXchange, FutureTravel that have shifted some or all of their call centre operations to homesourcing (or homeshoring). This is defined as "the transfer of service industry employment from offices to home-based employees with appropriate telephone and Internet facilities". Homesourcing is best thought of as a combination of outsourcing and telecommuting. In industry terms, such collectives are often called VSCs or Virtual Support Centres.

Thomas Freidman discussed this emerging trend in his book *The*

World is Flat: A Brief History of the Twenty-First Century (2005), giving support to the concept of telecommuting and changing workforce models. In 2006, a *Business Week* article entitled "Who's Helping Homeshore Workers?" referred to an IDC research that estimated that homeshoring was increasing at 20 per cent annually and was set to explode in developed economies. Well it hasn't quite exploded yet, but if staff turnover continues to be a challenging problem for organisations running contact centres, this could be a way of attracting new resources to the sector, particularly the onshore segment.

In terms of costs, homeshoring can be quite attractive for employers as there are no additional telecoms or office costs to incur. Most homesourced employees are independent contractors who use broadband connection to link themselves into the call centre system supported by the employing company. Such employees are typically very loyal because they value the opportunity to work from home and the flexibility it gives them. Moreover, by using their home as their workspace, they get considerable tax deductions. Homesourcing also enables the employment of disabled workers who cannot easily commute to the call centre, but who can otherwise be a very effective resource for the company.

The drawbacks include concerns over data security, monitoring of performance, control over staff, and the natural desire of people to work in teams. Companies that rely on homesourcing employ various strategies to combat such negatives, including managers who visit staff at their home to encourage them and keep them in contact with the organisation, virtual team meetings online, and instant messaging between team members for discussion and problem solving while they are on call. British Telecom reported in November 2009 that homesourcing was one of the new outsourcing trends that they were investing heavily in supporting.[10]

Customer Advocacy

Customer advocacy is an advanced form of customer service where companies focus on what is best for the customer, not just for the brand. It aims to build deeper customer relationships by earning new levels of trust and commitment, and by developing mutual dialogue and partnership with customers. Put simply, customer advocacy is doing what is best for the customer, even if that entails recommending a competitor's product.

To build customer advocacy capability, we have to deploy an additional layer of technology and an additional layer of support staff in the contact centre. The additional layer of technology captures customer contact history at every touchpoint—the Internet, call centre, branch—and is designed to ensure that every query or request is recorded in the system, so that at any point where a customer advocate sees an unresolved issue in the contact history, he can work with the customer to solve the problem. Now this does not always mean creating extra work for the bank. The job of the customer advocate is also to help empower the customer to resolve his own problems wherever possible.

Building an additional layer of support staff involves the deployment of facilitators between customers and the company. They are trained in cross-functional roles and empowered to provide customers with assistance in all areas of the business. The role of the customer advocate is threefold:

1. To be the main contact for the customer in handling a question or problem, and keeping the customer updated with timely and frequent reports as to the progress of resolving the issue.
2. To facilitate a resolution (ideally permanently) by bringing together the appropriate people within the business.
3. To implement a procedure that ensures the problem does not occur again, or recommend products or services to meet the customer's needs better.

A customer advocate also builds the concept of a "trusted advisor". That is, that the advocate will always side with the client in a decision. The idea is that if you truly provide the best advice for the client, even advising them to take a competitor's product, they will always return to you because you have established a trust relationship. Whereas if you are opportunistic and chase them only when you're trying to get more Assets under Management (AuM) or sell the next product, customers will become apathetic and even bitter.

A good example of the lack of customer advocacy in financial services is the case of my friend who enrolled with a global bank in their preferred banking scheme. He was required to deposit around US$150,000 to meet their criteria for a HNWI status and get the priority service that comes with holding that status within the bank, including dedicated branches, higher credit limits, plus a dedicated relationship manager. As my friend explains, he had happily been a client of this bank for about two and a half years, but he'd never once had a phone call or contact from his relationship manager in that entire time.

Well, at this time, he needed to liquidate some cash from his account to purchase a new investment property, and so for a while his account balance dropped below the minimum balance required for a customer of this "status". What happened? Within just a couple of weeks of the change in his account, he received a letter from the relationship manager telling him that he better deposit some more funds into the account or he would lose his preferred status! How's that for loyalty to the customer?

> *"Foolishly thinking you can speak to a human being and then when you finally do, foolishly believing they will speak the same language!"* Kim McNamara's response to the author's Facebook poll about dealing with banks these days

A customer advocate programme would prevent such communication from going out to clients, and would enable the bank to turn this into a positive opportunity, rather than a potential loss of customer. Incidentally, within a few months of receiving this letter, my friend withdrew almost all his funds from this bank and now doesn't use its services except for the credit card he maintains.

> *"When you make a request for a change to an account, etc and you're told that it's been done, changes should come through in a few hours/days ... but they never do. Calling up again and they have no record of the change request."* Troy King's response to the author's Facebook poll about dealing with banks these days

Tactical Channel Improvement

So what is on the contact centre improvement roadmap that we can achieve in the short term that will bring benefits to both the organisation and the customer? Over the next few years, think about the following initiatives set out in Table 4.2.

Table 4.2 Proposed roadmap for channel improvements

PROJECT/INITIATIVE	DESIRED OUTCOME
Staff retention programmes	Consider homesourcing as an option for retaining your best staff. Assign new cadets to the call centre for at least three months within their first year of service.
Full integration of email, VoIP and IM directly into the contact centre	Integrate the technologies customers use into the contact centre; picking up the phone is not a superior choice for customers. Don't forget to develop a communication policy for all employees that covers the use of the above, plus blogs, social networking sites and virtual worlds.
IVR menu redesign	Think about prioritising menu options based on traditional traffic analytics, thus reducing IVR navigation for those calls that are most frequently made. Incorporate voice-recognition "emotive" IVR technologies to redirect upset customers to a specialist "customer advocate" and defuse difficult situations.
Single-screen customer dashboard	Improve customer knowledge, process and workflow with a single-screen interface for CSRs, akin to the Internet banking centre for customers. Reduce the current workarounds with multiple disparate systems, separate logins, screens, etc.
Improved service culture	Work to create a total service culture within the bank that gives contact centre staff pride in their role within the customer equation, rather than feeling like they have been relegated to the call centre dungeon. Empower staff to solve problems, rather than creating a process that frustrates resolution through convoluted organisation structures.
Customer analytics	Use customer analytics to better understand the reasons for the call and work to anticipate customers' needs both collectively and individually. If a customer regularly calls with the same request, utilise analytics to serve up an IVR menu that prioritises those requests by CLID function.

Table 4.2 (cont'd) Proposed roadmap for channel improvements

Project/Initiative	Desired Outcome
Customer advocacy programme	Build a customer advocacy team that can act as a gateway between customers and the institution. Help this team to be champions for the customers and have them involved in any decision in respect of changes to customer process, new product launches, etc. that impact the frontline.
Dynamic offer management	Create a customer dynamics team (incorporating product specialists, marketing staff, customer advocates, etc.) to craft offers for segments that emerge from the customer analytics. Create tons of sales scripts and offers for unique, tailored cross-sell and up-sell opportunities that feel like better service for customers, rather than just a sales pitch.
Reform legal and compliance departments	Give the legal and compliance department KPIs for enabling customer solutions and improvements, so they are working for the customer and not just mitigating risk for the brand.

KEY LESSONS

Call or contact centres have been hailed by industry as a significant improvement in the ability to provide rolling support for customers in an increasingly mobile and time-poor environment.

IVR systems and channel migration have provided a significant cost savings imprimatur for corporates. This revolution, however, has not resulted in greater customer satisfaction. Increasingly, contact centres are experiencing problems with very high staff turnover, management are demanding improved sales results and corporations are grappling with the question of outsourcing versus onshoring.

The vision of the ultimate contact centre currently appears to be a convoluted mix of unified messaging platform, IP-based architecture, automated voice response systems and first-call resolution KPIs. *(cont'd)*

But do core building blocks still have to be put in place for this channel to be truly effective? This chapter tackled the issues of staff retention and effective measurement of the performance of the call centre. But it also looked at the deeper issues of the contact centre becoming the platform for all multichannel contacts with the customer—processing and recording contact history and optimising responses, whether through a sales opportunity or a better IVR design. The construction of a simplified contact centre dashboard or interface would reduce workload for CSRs, improve first-call resolution opportunities and result in better quality sales positioning.

Keywords Customer Advocacy, Up Your Service, Emotive Voice Response, IVR, Voice Recognition, Usability, Homesourcing, Outsourcing, Onshoring, Skype, VoIP, Instant Messaging or IM, Turnover, India, the Philippines, Sales Strategies, Analytics, Call Centre Dashboard, PABX Integration, Green, Social Responsibility, Contact Strategy, Communication Policy

Endnotes

1   Seventh Annual Report on Call Centres in the UK, Incomes Data Services, Call Centre Pay (Briefing 71)

2   "India's Call-Centre Jobs Go Begging", Sudhin Thanawala, Time.com. 16 October 2007

3   "New Trends in Performance Management 2010", PeopleStreme Research Whitepaper

4   Various sources: The Radicati Group, Wikipedia.org, Forrester Research

5   Workplace E-Mail and Instant Messaging Survey, 2004. The ePolicy Institute and American Management Association

6   IBM Virtual World Guidelines, IBM Research

7   World Trade Report 2009, European Centre for International Political Economy

8   Wikipedia article: "Instant Messenger"

9   "UK Call Centre Jobs Going Offshore for No Business Benefit", June 2007. Compass (www.compassmc.co.kr)

10 "British Telecom is up to its usual tricks", Anna Tims, *The Guardians*, 27 February 2009

5 Web—More than 10 Years Old ... and Still Broken

ALTHOUGH the Internet or "Web" has become an essential part of banking and commerce for most consumers in developed economies, the channel itself is still often limited in terms of potential because it is seen primarily as a cost saving mechanism by most retail institutions. There appears to be a widely held belief by many banking executives that while the Internet as a channel may supplement revenue, it is never going to be a serious sales or revenue channel. However, there are various facts that absolutely contradict that assertion, if not now, most certainly in the medium term. So the question is why do some make this assumption?

Many organisations that we've spoken to over time have argued that they've "tried the Web" and customers are just not using it as yet, and that it is a matter of market or user maturity. In developing economies, that may indeed be the case due to narrower adoption rates. However, in developed markets, if customers are not using the channel, it is more likely about channel effectiveness—how easy it is to engage on that channel, and having the right mix of services and products to ensure success. It is not normally a matter of poor adoption rates. There are various mechanisms that affect the ability of a retail bank or a financial services provider to generate revenue online consistently and profitably. The primary mechanisms are:

1. **Right product mix and journey.** Not all products work online, not all segments want the same products, and not all products should be dealt with in the same way. Getting the mix right is critical to online success.

2. **Findability.** Eighty per cent or more of users find websites through search engines,[1] so making sure customers can find

you and get to you through search engines is critical. In reality, integrated marketing campaigns do little to drive customers to the Web that behavioural shift doesn't already do.

3. **Usability.** According to research, usability problems are the lead causes of failures in realising revenue opportunities. (43 per cent of purchase attempts fail—Creative Good [source]; users find information only 42 per cent of the time—User Interface Design; companies lose 50 per cent of potential sales because users can't find what they look for—Forrester Research).

4. **Value Exchange/Content Scoring.** Ensure that the content you put on the site is both relevant and generates revenue or cost savings opportunities.

5. **Cross-sell to existing customers.** With increasingly complex compliance requirements on the front line, selling to existing customers is the easiest proposition online because you already have the customers' details. However, most retail banks offer no products to their Internet banking customers behind the "login" and see the platform as a transactional platform only.

To be honest, most of these mechanisms could be and often are dealt with in separate distinct publications dedicated to the specific topic. However, our objective in this chapter is to provide an overview primer of these critical pillars for bankers to get a holistic view of the right approach, and not rely on just one of these mechanisms alone

Figure 5.1 Mechanisms affecting online revenue generation

for success. In this way, a better rounded approach to Internet channel development should ensue. It also puts pressure on the organisation to think through where the channel fits strategically, on a product and customer basis, rather than having the luxury of assuming we can tack a product on to the channel independently, without impact on other channels, either positively or negatively. This chapter is one of the more comprehensive in the book for that reason, i.e. it is rather ambitious in its scope. Let's look at product mix as the first critical element.

The Right Mix

It is fairly obvious to all but the most passionate Internet evangelist that certain products just aren't going to work on the Web. There are various reasons for this, but the three most critical drivers for online success are convenience, control and complexity (or conversely, simplicity).

Convenience is a primary driver in online adoption, with 75–96 per cent of users consistently giving their primary reason for going online to apply as either time saving or convenience.[2] Convenience simply means it must be quicker and easier for a customer to get this online, compared with picking up the phone or going down to the branch.

Control is a driver online because the user has complete control over the engagement and is empowered by the channel to do it himself. For tech savvy early adopters, they get great satisfaction out of achieving success in this manner and this fits with our Maslow's model of self-actualisation discussed in Chapter 1.

The last driver is **complexity.** If a customer has to spend 45 minutes online filling in forms, only to have the application process potentially fail at the last moment because he doesn't live in the right location or doesn't have the right details on hand, it is pretty frustrating. For customers, the online process has to be easier and simpler than what they are used to in the real world. This last element also leads strongly into the field of usability, and we'll discuss this later in the chapter in more detail.

In Chapter 2 we identified the types of products that customers typically look for online. They included various products, but generally they are simple products that require zero or little interaction with a customer service officer or teller in the execution phase (Table 5.1 overleaf). That is, they do not require the assistance of an advisor to help them with the execution of the application or placement of the order, and there is no upside to going

Table 5.1 Preference for retail banking products online (by market)[3]

INDIA	SINGAPORE	HONG KONG	UNITED ARAB EMIRATES
Credit card	Time deposit and monthly instalment plan	Time deposit and monthly instalment plan	Credit card
General Insurance	New bank account	Stocks and company shares	Personal loans
Bonds, unit trusts or mutual funds	Credit card	Credit card	New bank account
New bank account	Foreign exchange and currency transactions	Foreign exchange and currency transactions	Car loans
Personal loan	Bonds, unit trusts or mutual funds	General insurance	Stocks and company shares
Stocks and company shares	Stocks and company shares	New bank account	Time deposit
Time deposit and monthly instalment plan	Fixed income products	Fixed income products	Monthly instalment plans

into the branch to fill out the related form. For investment class products, we generally accept that an advisory role is critical in the pre-purchase phase, but we increasingly see the commoditisation of execution where this can be enabled online. In respect of other more complex products, again the advisory role is critical to the selection of the product, but education of the customer and capturing his interest can be successfully supplemented by the online channel.

As products increase in complexity, it is less likely that the entire process of purchase or application can be done through online fulfilment, and indeed due to compliance rules and regulations, it is often simply impossible. There are, however, parts of the product selection process or purchase mechanics that can be assisted through the online channel. For example, a bank can educate customers on the benefits or performance of a product so that it can assist customers to select the right product to meet their needs. A financial services company may utilise the Web channel as a lead generation tool, capturing interest in the product and executing offline. Customers may use the channel in the maintenance phase to renew a policy or redraw a loan, or simply to track how their investments are performing. So as a tool, online purchase or the actual sale is not necessarily the ultimate goal for a product on the channel.

In simplistic terms, there are three main phases in respect of the customer-product relationship that we can encapsulate through the online channel in varying levels of sophistication. They are:

- **Pre-Purchase or Product Selection.** What marketers might typically call the "research" or selection phase. This is where customers seek to be informed on options available, choose their budget or quality options, etc. Increasingly the Internet is a primary element in pre-purchase selection.
- **Purchase or Execution.** What the bankers call "acquisition"— the moment the customer selects the product and agrees to the purchase or fills in and submits the application form.
- **Post-Purchase or Maintenance.** In finance parlance this is where customers monitor their investment products, or where they make a claim on their insurance product, or where they check their statements. It's everything that happens after they sign up for the relationship with the bank via a specific product.

Not all retail banking products are created equal in respect of their suitability to the online channel. Primarily this is due to the sophistication or complexity of the product. Products can be broadly grouped into the following categories:

- **Simple products.** No advice needed. Examples are credit card, current/savings account, personal loan, general insurance.
- **Complex products.** Advice needed. Examples include mortgage, life insurance, overdrafts.
- **Investment class products.** Specialist advice required. Examples are securities, investment funds, mutual funds, derivatives, structured products.

Figure 5.2 Typical phases of customer engagement online

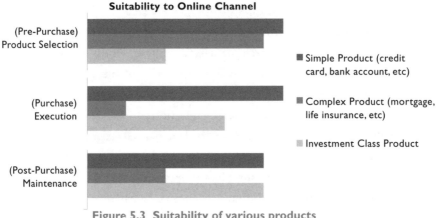

Figure 5.3 Suitability of various products
for the online channel at different stages

Putting this together, we can determine through analytics, customer behavioural analysis and customer research, a typical pattern of utilisation or suitability of the online channel in supporting each respective phase of engagement and for each class of product. Let's analyse this in greater detail and look at some examples.

To illustrate the issue of online engagement, let's take the example of a mortgage product. Now, subprime aside, there has been a significant increase in the mortgage business over the last few years globally, and a borrower is definitely finding it easier nowadays to get access to mortgages than he did, say 20 years ago. Many bankers, however, might dismiss mortgages as too complex to sell online today. The facts show that such a dismissal would be unwise. Although the final process of the application for a mortgage may be done face to face, the entire process is often handled online or online with the help of the call centre. In fact, Countrywide from 2006 onwards topped the list of mortgage providers in the US, and as early as 2001, a full 85 per cent[4] of its business came through the online channel, including those through wholesale mortgage brokers.

So, the largest seller of mortgages in the US has done almost all its business through its website over the last seven to eight years, but most banks don't even generate mortgage leads effectively through the online channel. What is wrong with this picture?

Let us assume that the Countrywide experience shows it is entirely reasonable to sell mortgage products through the Internet channel. What is holding back the majority of traditional banks from executing this successfully?

There are normally two key issues. The first is purely a lack of experience in constructing the proposition for the online channel, no customisation of the application process, and a poorly designed interface. In other words, usability and customer experience. The second issue is inflexible compliance requirements, not from a regulator's perspective, but from the internal legal and compliance departments of the bank. Additionally, many banks simply aren't having any success with this product online because the only choice a customer has is to walk into a branch or call the centre.

In many cases, the legal and compliance departments are the institution's own worst enemy when it comes to revenue generation through alternative channels. Application processes that have been branch-tailored, based on the concept of a defined legal process with strict terms and conditions on a paper application form, rarely fit cleanly online. When compliance departments are confronted with an exuberant Internet team that wishes to sell online, the normal reaction is a resistance to anything that isn't effectively a complete replication of the physical "branch" legal process. This makes innovation of the application process online extremely problematic. As the HSBC travel insurance case study at the end of this chapter shows, by adapting a product to online properly, the rewards can be significant. Pure top-line revenue improvements without any negative impact on the branch are possible by creating the right environment for customers to interact.

Findability

Google revolutionised search engine behaviour in the late 90s when they launched their portal of the same name on the unsuspecting public. Prior to that, we had lived quite happily with the equivalent of a yellow pages or Seers mail order catalogue system online. With the launch of Google.com, we learned that finding stuff on the Web could be really, really easy.

Prior to the revolution in search engine technologies and interface, we relied on URL marketing through traditional media, banner add click-through and third-party link population to get people (or eyeballs, as we called them in dot.com speak) to come to our site. But as early as 1998, Georgia Tech reported in their annual World Wide Web survey that 84.8 per cent[5] of users found websites using search engines. This was even BG—Before Google. So imagine what those stats look like today! The question of course is how this translates to real revenue.

NPD Group, a research group specialising in consumer purchasing and behaviour, tested the impact of the various mechanisms that could drive a consumer online and which were most effective in initiating a purchase. The results showed that 55 per cent of online purchases originated from Search listings (Figure 5.4).[6]

In the midst of the global financial crisis, the Web was a haven of activity, again bucking the trends of doom and gloom in consumer sales. US consumer technology e-commerce sales increased 19 per cent to more than $700 million for the first two weeks of the holiday season compared with the same time last year.[7] This trend is consistent with search engine marketing for financial services products also, as can be seen by Pay-per-Click (PPC) activity as reported by Microsoft and others. As Microsoft's Craig Brown reports:

> "As the general public becomes more concerned with their financial security, adCenter accounts in the banking sector have seen average YoY increases of over 200% in impressions and 190% in clicks (September '07 vs. September '08). These increases have been steady since January '07."[8]

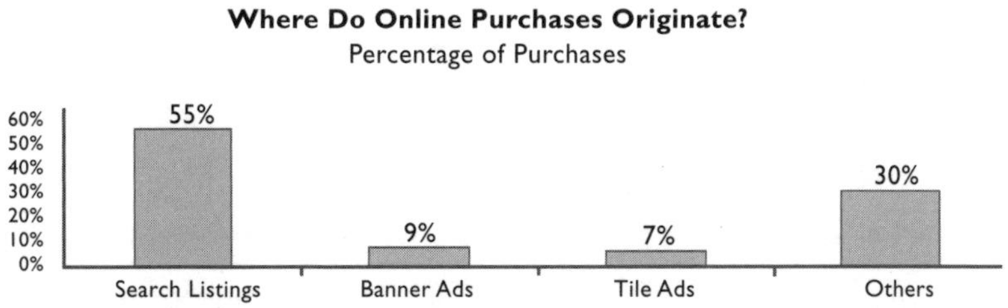

Where Do Online Purchases Originate?
Percentage of Purchases

Figure 5.4 **Origination of online purchases** (NPD Group Survey)

So let us summarise. Statistics show that if you want to move customers online because of the significant channel cost saving benefits or for the purpose of capturing new revenue, you need to start by optimising your online products for search engines, or making sure you have the right presence on the right search engines. That is going to account for somewhere between 55–85 per cent of your potential traffic to your product pages.

What are the options in respect of helping consumers find us through our search engine of choice? There are two mechanisms that improve your visibility, or findability, on search engines.

The first method is broadly categorised as **Search Engine Marketing** and the second is **Organic Search Engine Optimisation.** The first can pay dividends at any time and can be used for specific campaigns or for specific periods of time, but is pay per use, just like traditional media. The second is a longer-term strategy that costs upfront, but effectively produces longer-term results that cost much less to maintain. To make it simple, let's review it from a results page perspective from Google.com (Figure 5.5 overleaf).

For short-term advertising campaigns then, pay-per-click SEM (Search Engine Marketing) is probably the most cost effective method. Google is the largest search engine in the world with between 49–60 per cent of search engine traffic in the US,[9] depending on which comparison study you select. Google uses a system called Adwords (adwords.google.com) for customers to register and bid for keywords.

Adwords and other such systems are extremely comprehensive and allow geographical bids, as well as campaigns for very specific time periods, day by day, in fact. Yahoo Search Marketing (previously known as Overture.

Figure 5.5 Google search results as they pertain to organic vs pay-per-click strategies

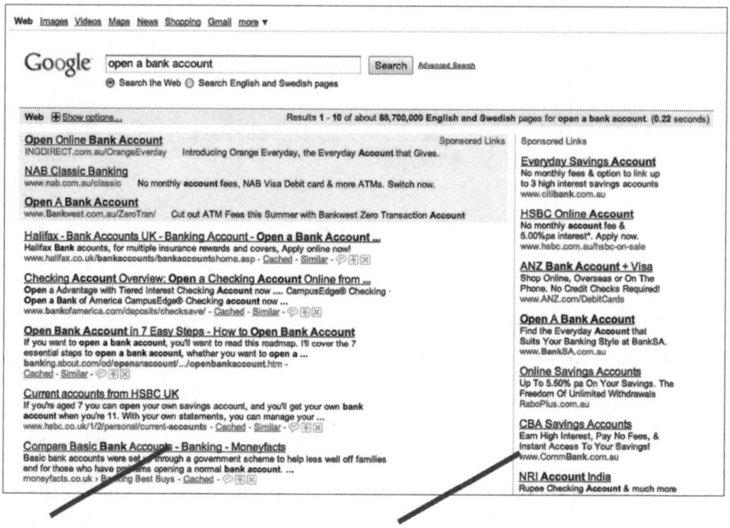

ORGANIC SEARCH results appear in the main body of the search results page and cost nothing per click, but require you to redesign your website content/source code and improve your site's link popularity. It normally takes about 3 months to see results.

SEARCH ENGINE MARKETING results appear in the "sponsored link" section of the Google search results. Essentially, you bid for the keywords you want, and you pay at the successful bid rate every time a user/visitor clicks on your link. The results appear immediately, but only stay for as long as you are prepared to keep paying.

com) offers a similar service, as does MSN's AdCentre (adcenter.microsoft. com). In fact, Google's Adwords is largely based on Overture and was the subject of a US Federal Court action.

The more popular a keyword, the more expensive it will be per click. For terms such as "Class Action Lawsuit", you might pay US$30–$40 per click. As of January 2008, "Ortho Evra Lawyer", keywords related to a class action against a once-a-week birth control patch, were the highest bid keywords. During 2006–07, keywords on Google (US) such as "Leads Mortgage" would have cost you US$30 per click, "Credit Card Debt Consolidation" US$25 per click, "Equity Home Loan" US$15, and "Student Credit Card" US$10 per click. So in the case of "Student Credit Card", you could pay US$10 every time a visitor to Google clicks on your sponsored link! So a popular search term might be costing you US$3,000–$4,000 a day in click-thru traffic. For smaller country sites such as Hong Kong or Singapore, the fee may be significantly lower (say US$1–$3 per click), but you still pay based on volume; there is no cap.

Alternatively, if you want long-term exposure for a product or your brand attached to specific keywords, then you have to **optimise** your website to ensure that Google, Yahoo and MSN index your site according to those select keywords. That means that in the content on your Web pages, you need those keywords to occur in multiple locations (page title, headings, bold font, hyperlinks, main text, etc.) and within source code so-called "meta-tags". This is usually a one-off investment to revamp your site to be more effective, but the results pay off for a longer period, usually taking about three months to take full effect and lasting for up to a couple of years of quality exposure. Which costs more?

Upfront, organic SEO (Search Engine Optimisation) is more costly, but there are no or minor ongoing costs. Whereas pay-per-click and SEM costs you every time you want to use it and it stops working for you when you stop paying. The fact is, redesigning and rebuilding your website to make it more effective for search engine indexing is a good long-term investment in any case, as the Web is more likely to become a more important marketing tool for your brand as we move forward in time.

Usability

Usability is a term we use to describe how easy a system, product or interface is for individuals (users) to use. In recent times, the term has come to include a broad range of references, including the art of user-centred design (UCD), interaction design (IxD), usability testing (UT) and other such fields of endeavour. The primary notion of usability is that an object designed with the user's psychology and physiology in mind would be, for example:

- more efficient to use (it takes less time to accomplish a particular task);
- easier to learn (just by observing the object); and
- more satisfying to use.

By applying this particularly to the engagement process for a customer seeking a banking product through an electronic interface, such as Web, ATM, mobile phone or branch kiosk, we can improve the closure rate of new product applications and improve the return on investment spent in the underlying technologies. To illustrate a simple example of how interfaces are designed without any thought as to their usability, or even adapting a process from the physical to online, review the instructions from an online credit card application form for a retail bank in the UAE (Figure 5.6).

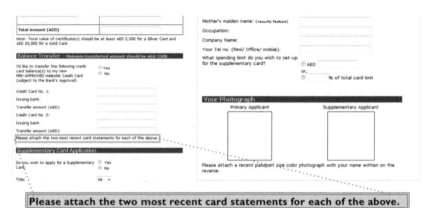

Please attach the two most recent card statements for each of the above.

Figure 5.6 Adaptation of an online form that obviously failed

In at least two locations on the online form, customers are asked to "attach" a physical document to the form. This might seem ludicrous (do I staple them to the screen; won't that damage it?), but these sorts of errors creep into the system where physical application forms are simply re-tasked for online without any thought given as to how the user might comply with the request. **This is a design error.**

In other instances, we see customers being asked to print off application forms and take them into a branch. **This is a usability problem** because instead of simplifying the process for applying for a product, we've just increased the complexity by introducing both the step of printing the form and the requirement to visit the branch physically. We've completely missed the intent of the Internet from a user perspective, which is *to save time*. It would be better not to have the application form online at all in this respect.

Again in the interest of not rehashing all of the very good publications that are available on the subject of usability, let's instead review a few simple steps that banks can use to improve usability on specific product lines.

Watch your customers (users) in action

Just to be clear, we're not talking about stalking customers here, but the first thing we need to establish is where the problems exist in the current system. Usability testing was first established back in the early 1980s by Xerox, and was becoming popular as "Human-Computer Interaction" within the better MIS schools of leading universities. Leading usability pundits such as Jacob Nielsen and Don Norman started to describe the process of usability testing as the Internet emerged commercially in 1994.

Usability tests can be done in a variety of ways. Typically, an observer sits and watches a user go through the process of using a particular device or interface, and records her observations about the various issues the user faces with the task. Usability tests can be free-form, where the user essentially just "plays" with the interface, or can be controlled, where the user is given a specific list or set of tasks to accomplish. The observer is just that, a silent observer.

Figure 5.7 Informal observation of a usability session (Credit: UserStrategy)

The observer does not actively get involved in instructing the user on how to carry out a task, as that could bias the results. The observer's job is to observe and take critical notes on areas for improvements. As such, observational data can be hugely invaluable. When we were usability testing the HSBC Personal Internet Banking portal in 2002, we asked customers to change their email address in the system. Only 10 per cent of subjects could complete the task. Observations showed that they did not know where to click on the site.

The Internet banking portal for HSBC at the time contained a number of areas to edit, among other things, your personal details, Internet banking settings and limits. The menu item was named "Services". Now from an IT perspective or perhaps an internal bank perspective, this makes sense, but to customers … not a hope. The observational approach gleaned this gem. When we changed the menu item to "My Details", the success rate went from 10 per cent to 100 per cent.

The observational approach

So what core elements of your multichannel service environment should you be observing and testing frequently (at least once a year)? You need to look at these:

- Public website;
- Secure website or Internet banking portal(s);
- ATM;
- In-branch systems (good idea to test both teller systems and those available to the public at large);
- IVR or automated call centre systems; and
- Mobile applications and/or widgets (desktop components).

Observational field studies are another type of usability test, where we simply observe customers coming into our branches or Electronic Banking Centre (ATM) locations and watch the way they interact with the environment (as we noted in Chapter 3 in relation to branch improvements). This is difficult to do for Internet banking though, as we can't just invite ourselves into customers' homes and offices to observe them. Thus, it is more likely that usability tests would be carried out in a formalised manner.

If we assume that we are going to observe our customers, the question then emerges as to what we are looking for and how we are going to use that information. The first thing we are looking for is purely problems in

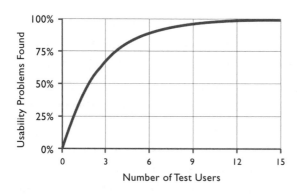

Figure 5.8 Nielsen's extrapolation of usability problems found per number of users

usage—areas where customers can't intuitively use the system or take too long to complete a task. The observer can ask some simple questions such as "What are you thinking?" or "Can you describe for me what you are doing at the moment?" Bad questions would be "Do you like this…" or "Do you need any help?" The observer needs to stay completely objective in this sense.

Testing with just four or five customers can typically return 80 per cent of the issues with a particular system or interface,[10] so this need not be an exorbitantly expensive process. Here are the key issues to look for in a usability test:

- User-centric language instead of bank-centric language.
- Flow problems where users have difficulty understanding the process.
- Form or page design, where users don't understand where to look.
- Understanding the purpose of the page or interface.
- Successful completion of the task or objective.

Put the customer element into the design process

Improper language and poor design of the interface are the most common issues facing banks in respect of usability. Why? This occurs mainly because bankers and IT staff design the interface or systems with absolutely no involvement from customers at any stage of the process. How well do your IT staff actually know your customers? The reality is that IT "guys" are probably the last people you want designing interfaces that customers are going to use on a regular basis.

Improvements in interface design have come rapidly in the last few years. Organisations such as Apple, Google and others have put huge emphasis on making systems simple to use, and they have been rewarded with record sales and cult-like followings. Again, such processes need not be expensive; indeed, employing interactive design methods will probably save you development costs for a system.

Unlike traditional interface design which takes place entirely behind the scenes and without any reference to customer usage, the objective in interactive design is to use low-fidelity (or lo-fi) methods of prototyping and multiple iterations to get to an optimal interface design. These methods are designated as lo-fi because they don't require any special technology, software or skills. In fact, the most common methods of design in a lo-fi environment are simply pen and paper. We can test off these lo-fi prototypes also, getting input from customers before we've even started coding.

Let's take an interface for a credit card landing page within an Internet banking portal. The first thing to do is to think about who the audience is, and what the objectives, metrics or success measures of the page might be (Table 5.2). Then we need to work up a sketch of what the page might look like, refining the design through exposing it to both customers and key product/business stakeholders.

Table 5.2 Credit Card Landing Page

INTENDED AUDIENCE/NEED	METRICS
New Customers (Application)	# of new card applications (acquisitions)
Existing Customers (Payment)	# of new direct debit authorisations
Existing Customers (Usage)	# of redemptions (bonus), offer take-ups

Stage 1: Informal whiteboard sketch

Brainstorming of the key content elements that meet the needs of the intended audience and meet the required metrics of the bank can be done over a whiteboard or a sketch pad. You can include credit card product specialists, customer advocates from the call centre or branch, and even staff who are customers in these exercises. IT guys are not required.

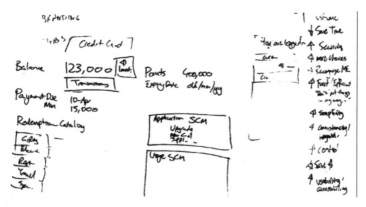

Figure 5.9 **Informal whiteboard sketch**

Stage 2: Formal sketch (lo-fi) ready for usability test

The beauty of the lo-fi approach is that it can be done with very basic tools—a pen and paper, whiteboard, butcher's paper, anything where you can quickly sketch the ideas that flow from an interactive design session. It can be formalised by transferring this into PowerPoint, Visio or Photoshop, but the objective here is speed and testing.

How can you test a paper prototype with real customers? Easy, give them a pen or pencil and ask them to mimic a mouse and show you where they would click for a specific task. Customers can easily make the conceptual leap to visualise how this would work in a live/real setting.

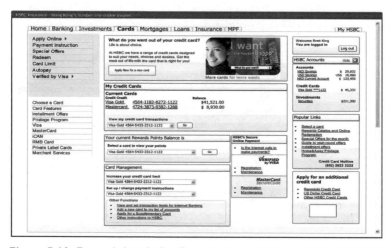

Figure 5.10 **Formal sketch (lo-fi) ready for usability test** (Credit: HSBC)

Stage 3: Creative concept and implementation

Further usability testing can be done at each stage, and it's not too late to trial this in front of customers at this point. Remember though that the earlier you test, the more accurate the screen designs will be at the end. The later you leave the testing, the more likely it is that you will have to compromise between a key marketing concept or design element, and something that actually works from an interaction perspective.

The problem is that marketers and designers generally don't understand the interactivity or the dialogue-like elements of the "interface", and IT guys just don't get access to customers on a regular basis. So, in the typical design process you are having two groups fight for the "right" to determine which way the screen design should go, when neither is adequately qualified to be the voice of the customer.

If you want to get your Web stuff right, you must have customer advocates as part of the design process as a minimum, and ideally, real customers. Remember, staff are real customers too (read inexpensive), so they make an excellent source of usability test subjects. But in the end, nothing beats putting real customers into the design and testing process.

To get started, pick up a Usability or Interactive Design book or kit to get acquainted with the process and figure out who in your team should be taking the lead on this.[11] Most large institutions now have their own full-time usability teams. However, the one caution most usability practitioners offer is to take the politics and pet projects out of the design process.

Figure 5.11 Creative concept based on successive wireframes
(Credit: HSBC)

Value Exchange and Content Scoring

This gets to the real meat of the conversation about what the institution's website is for. If you asked the management of the bank, you might get a range of answers including:

- promoting the bank and offering the latest news;
- giving customers access to the bank's products, services and special deals;
- telling customers, investors and others about the bank;
- delivering financial reports and company information;
- offering the latest marketing campaigns and offers; and
- providing links to other sites within the group.

The problem with much of this is that the information is simply not relevant to the vast majority of visitors to your website. Most bankers probably imagine that there are a whole range of stakeholders such as investors, journalists, corporate customers, retail customers and students coming to visit the bank's website daily looking for vast reservoirs of information. The reality is that more than 90 per cent of your daily visitors to your website are retail customers looking for retail products and services, and the others rarely visit.[12]

Unfortunately, the website is one of those properties that everyone in the bank claims ownership of, so every time it comes up for redesign it is the perennial "bun fight" over who gets which part of what is perceived as the most valuable real estate, i.e. the homepage. If you challenge the CEO of a bank to define what should be on the website, he'll probably opt for financial information and latest PR news because that is what he wants the market and investors to hear.

The premise of a good website, however, is really providing value to those individuals coming to visit the site most frequently. The data simply show that journalists and investors rarely visit your site, except perhaps once a quarter when you release your financial results. Even at those times, the percentage of visitors who are NOT retail customers is less than 1 per cent. This introduces us to the underlying concept I like to call the "value exchange".

The value exchange

The principle behind the value exchange is pretty simple. You cannot hope for customers to come to and utilise your site, unless you provide them with some fundamental value. Saving the institution money and making your revenue numbers look better is NOT a key goal of your customers.

So what is? Simply put, customers want you to improve their lot in life, respect their increasingly valuable time and save themselves money. Preferably doing all of this together at the same time. Big demand on a bank, right? Not really. The fundamentals of a loyalty programme are at work here. That is, customers who make the most margin for you should be valued more highly and should be recognised and categorised as such. This is, after all, why we have brands such as CitiGold, HSBC Premier, Wachovia Wealth Management, Standard Chartered Priority, ABN Amro Preferred. We recognise that segments of customers are "valuable" and need specific services.

Let's assume that 70–80 per cent of your visitors to your website are retail customers, and about half of those are HNWIs (High Net Worth Individuals). If those individuals were to walk into your wealth management dedicated branch, would you insist they read the five latest company press releases, a message from the CEO on the latest financial report and watch the latest TV commercial before you allowed them to speak to a relationship manager? No way. But guess what? That's exactly what many banks are still doing today with their websites. Why? Because they don't understand the

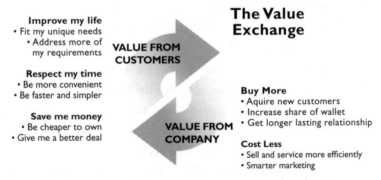

Figure 5.12 The Value Exchange concept as it relates to the drivers for an effective Web presence

audience that actually utilises the site on a daily basis. Simply put, they are more concerned about telling everyone what they have to say rather than helping actual visitors get to where they need to go. Let's look at the various categories of bank sites and how the typical site stacks up.

Brand portal websites

Figure 5.13 shows the websites for Standard Chartered, HSBC Banking Group and RBS as of late 2009. They represent a selection of some of the top banks in the world from different regions, but they share one thing in common, that is, they have either a large network of branches or sub-brands with their own identities in separate markets.

In these instances, what is the objective of the portal site? In most marketing, advertising, printed collateral or "press" that a bank releases today, they will use a primary URL (Uniform Resource Locator) or Web address. Typically that address is **www.nameofbank.com**. This is an easily identifiable Web address that customers can find quickly. Indeed, when customers type your bank name or brand (e.g. HSBC) into a Google search engine, this is the most likely website they will find at the top of the search engine listings. So if I am a RBS customer and I have forgotten the local country website domain for my local RBS bank, I simply go to Google or Bing and type in "RBS".

Figure 5.13 Examples of brand portal websites

So what does a portal site need to do at this point? Its objective is really to get the respective visitor to the section or website of the bank that is most relevant to them. Call it a gateway, or traffic sign, that gets you to the part of the bank you are most interested in. This is different for country websites or smaller brands because for those sites you need to get straight to the sales message. For bigger brands, it is about getting you to your destination as quickly as possible. The site should also be easily identifiable as your core, corporate brand website. It should show your core values as a brand.

Customer-focused websites (typically country sites)

This is the more common type of banking website you will find. It's basically a sales and marketing platform aimed at customers of the brand. Typically, we find these at a country or brand level, specific to a location or segment of customers. These customer-focused sites are intended for all customers within a brand, or there might be different versions for different segments of customers, for example, a "personal" banking site, a wealth management division site, and a site for SME customers.

More often than not, the country site focuses on all customer segments and visitors. The typical URL is **www.nameofbank.com.country,** where country is the two-letter country domain identifier for your location.

The problem with the homepage of this site is that everyone—the product team and every department within the bank—thinks this is the place where they absolutely must put their message to their particular stakeholders. Thus it can get pretty confusing trying to manage the design process and negotiate internally within the bank as to the end result. Sometimes it's simply impossible.

The goal should be simple however. This site is a portal or directory of products and services that **targets customers first,** and is focused on driving revenue results, with corporate communications a low priority.

Here are the websites respectively for Bank of America (BofA), Citibank US, DBS Singapore and Hang Seng (Hong Kong). While they all vary slightly in approach, they all do a pretty good job catering to the core needs of their customers. Hang Seng (Hong Kong) and Citibank (US), however, stand out above their contemporaries because clearly some research into customer needs has taken place. This can be seen in the action-oriented and revenue sensitive navigation elements visible on the homepage.

So which will work better for your institution? The facts show that the customer focused sites typically get 30–50 per cent more traffic than the portal sites,[13] so on the face of it, a product-led marketing approach is more profitable. However, for banks with more complex businesses and diverse marketplaces, just one single customer site doesn't work because a "one-size-fits-all" approach is impractical.

Figure 5.14 Examples of country sites

Content scoring

One vital step in the planning process for a revamp or design of your Internet capability is to establish a comprehensive list of content and/or functionality required for the revamped site or portal. The key objectives are:

- Inventory and analysis of **current content and functionality.**
- Review and **eliminate redundant or conflicting content.**
- **Prioritise content** for future site redesign and/or maintenance.

The data utilised in constructing the content audit and the consolidated list of content and functional elements are shown in Figure 5.15. These elements are then consolidated into a master "content" list which is then referred to the business and a sample set of customers for scoring and prioritisation.

The content scoring exercise is designed to establish a master list of content required for the site, prioritise that list according to both the needs of the customer and the needs of the business, and align business requirements, competitive analysis, call centre statistical analysis and extensive usability testing.

The result of this scoring exercise is a clear picture of the content

Figure 5.15 Data for constructing content audit, list content and functional elements

required for the website for the current revamp and possible enhancements to future iterations of the website as technical infrastructure and capability become available.

This provides a structured approach to content prioritisation rather than the typical internal department free-for-all that normally revolves around who has more political pull in the institution instead of whether the content is right for the site or not. It also identifies key elements that are currently not supported by the business requirements, but are high-priority items for customers. Thus, it's about getting the important stuff back on the site.

Scoring mechanism and resultant prioritisation

Potential content identified in the audit and research is then scored according to three components:

- **Customer needs (50 per cent weighting)**
  - ○ short-term impact of content on the customer
  - ○ current issues related to process (touchpoint)
  - ○ content value-added to the customer (long-term)
- **Business requirements (35 per cent weighting)**
  - ○ ability of content to save organisational costs
  - ○ ability of content to generate new revenues
- **Competitive analysis (15 per cent weighting)**
  - ○ Do our competitors have it?
  - ○ Is this a best practice feature (for reward programmes and in general)?

We score the content and create a prioritisation based on a mean score of between 1 and 5. Five (5) indicates a high-priority in all segments, whereas a one (1) indicates a low priority. On this basis, content is classified and prioritised in a structured manner rather than by some arbitrary decision of marketing or IT personnel.

3.0–5.0	High priority content (must be on the site)
2.5–3.0	Medium priority content (can be on the site)
1.0–2.5	Low priority content (shouldn't be on the site)

In many cases, the content or functional element scored is actually a problem identified or a solution to a current usability issue. In such instances, the solution will be modelled in the interaction design workshops.

Outcomes of content prioritisation

The outcomes of the content prioritisation are to:

- identify and confirm primary and secondary scenarios;
- identify key elements for navigation (card sort testing);
- identify strong candidates for inclusion on the site that have previously been excluded or were not technically feasible; and
- identify possible segmentation of non-air awards for search functions or cataloguing.

The items in Table 5.3 would be considered mandatory inclusions on the site because they scored highly in all three scoring elements, namely customer needs, business needs and competitive threat. Some of these elements may not be able to be supported by process or technology in the current iteration, perhaps due to budget or technology constraints. The inclusion of such elements indicates that they should be reviewed by the business carefully before they are excluded in their entirety.

Cross-Sell to Existing Customers

There is a very simple, but extremely valuable tip that I'm going to give you here that has the potential to generate millions of dollars in revenue for you over the next three to five years. It is so simple that you will kick yourself for not recognising this already, and it is so fundamental to the use of your website that it should drive your total budget decision and marketing approach from this day onward. There may still be some resistance from those who prefer the status quo, but when you check out the analytics of your own site, the evidence will be overwhelming, I assure you.

Here it is. Look at the website on page 174 and tell me where more than **90 per cent of daily visitors** click on this site.

- Is it the "Revolving Credit Facility" billboard-style marketing advertisement?
- Is it one of the other hot offers on display?
- Is it the "September Special" offers on the right hand side?

Table 5.3 Sample scoring matrix. High priority content elements Score of 3.0 to 5.0 (HSBC MPF website)

Content or Functional Element	Description/ Customer Needs	Total Customer Score	Total Company Score	Overall Rating
Employee termination process	Check whether employers have notified us of their termination, paid last contribution, etc	4.74	4.69	4.69
e-form	Download critical e-forms related to fund allocation and other updates	4.14	4.84	4.33
Frequently Asked Questions	Categorise the FAQs so that customers can find the questions easier; suggested categories include search function in FAQs section	3.92	4.50	4.23
Advanced search engine for MPF rules and options	Allows users to ask questions in simple, everyday language on various topics	4.44	4.38	4.21
Improve fund information (fund charts were separate on the competitor analysis as a 2.0, but is rated under here)	Provide daily unit prices for each fund, not only the latest unit prices. Search function on unit prices (key in date) Ranking in HKIFA survey	3.76	4.67	4.17
How-to procedures, based on high-frequency calls to the call centre	• Don't know how to fill in the forms • Payment procedures • When do I receive my payment? • Where to obtain advice on funds?	3.80	4.84	4.01

NOTE: The competitive score column has been removed for ease of reading.

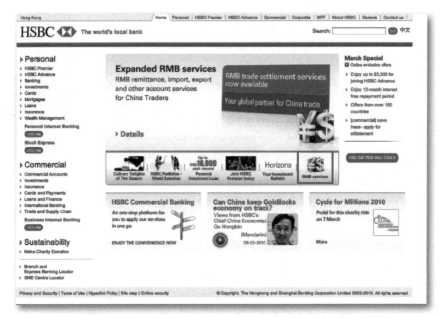

Figure 5.16 Typical retail bank homepage (Credit: HSBC)

• Or perhaps the search box in the top right hand corner?

Give up? More than 90 per cent[14] of customers visiting this website click on one section of the site. It is, of course, the Personal Internet Banking **login button.**

Now, most banks consider the "Internet Banking" portal, or "secure-site" as we sometimes call it, a functional platform for transaction capability. The focus is mainly on account balance, fund transfer, bill payment, term deposits, but it is normally run by the IT team who are the "functional" guys.

The fact is, today your marketing team is probably spending 95 per cent of their budget related to Web marketing on either building public websites that pitch product, or launching new campaigns on say, third-party sites. The fact is, based on these analytics, you need to be spending at least 90 per cent of your Web marketing budget on building offers and campaigns for **existing customers through the Internet banking secure portal.**

The other advantage to this is that the acquisition process is dead easy.

You already have all the customer information, so compliance is simply a click-based existing customer acquisition, rather than copious forms or entry to provide proof of who they are, their credit risk assessment, etc. These are simply the easiest customers to acquire.

A strong word of warning, however. Don't opt in for a simple banner ad approach here as all you will do is upset your most valuable audience. You have to think about this and provide relevant offers. So don't pitch a Gold credit card to someone who already has a platinum Visa card. Don't pitch term deposits to a customer who is already a Premier account holder with a managed fund. Don't pitch retirement plans to a student. For this you need business intelligence and segmentation that create compelling, targeted offers that appeal. There are some other issues though.

Combating banner blindness

In a fairly brilliant piece of early usability testing, Jan Benway and David Lane of Rice University in Houston, Texas, US, discovered in 1998 that users were starting to filter out "advertising banners".

The ad agencies that were still thinking of the Web as just another channel to push traditional media campaigns had shot themselves in the foot a little by producing little magazine ads and billboards everywhere. We just retrofitted these and called them banner ads. Let me explain.

In October 1994, *HotWired* (*Wired Magazine*'s former online brand) made history by placing on their website the very first banner ad. It looked

Figure 5.17 The very first banner ad appeared on *HotWired* in 1994

like this.

The ad was produced for AT&T by Modem Media and TANGENT (which went on to become part of Razorfish). Initially, banner ads were hugely popular and all through the dot.com age we were talking about banner ad conversion rates, click through rates, eyeballs and other such metrics which were very exciting ways to measure this huge new phenomenon. In 1996 and 97 when banners became the next best thing, all the big brand guys were getting online experimenting with this cheap, but very effective medium. Response rates on banners in these early days were better than any responses on existing media offerings. Perhaps it was because of the novelty value.

But as advertising agencies rushed to put more and more ads on this new medium, we lost the advantage the new platform gave us. You see, the Internet and banner ads could have been something fundamentally different. Why?

The Web, compared with print media, TVCs (TV Commercials) and billboards, is a very different medium. It provides the ability to interact, to engage, to have a dialogue. Traditional media did not allow this unique capability. However, ad agencies which flocked to the Web lacked fundamental creativity in adapting to this space because they were caught up in the concept of the message being the all important element. **The Web is about experience,** not simply a message. Therefore, by flooding it with banners that were static and simply duplicating what we were already doing with other media, we missed the opportunity to capitalise fully on this new medium.

In late 1997, some ad agencies started to experiment with rich media in banners, but their objective was really akin to trying to create something like a mini TVC in a banner ad, again not an experience, but the concept of a "rich message". They were myopically focused on the concept of trying to create brand recall, and not thinking about how to engage in a dialogue with the user, which the Web enabled. This was the fundamental shortfall in the early attempts to utilise banner advertisements. The same happened with Flash introduction pages, which were originally argued by traditionalist marketers as a great way to introduce the Web equivalent of TVCs. The idea was that I'll force you to view a 15-second TVC before I allow you to get access to the content on the website—a serious mistake.

As a result of our simply presenting more and more of the same up to customers who were increasingly being bombarded with much more "noise" across the media spectrum, it became clear that banner ads were simply being dismissed by customers as just more "noise".

If you are going to serve up banner ads, you now have to retrain your customers into thinking that these are relevant, timely and appropriate. So start figuring out how to identify your customers and serve them content that is tailor-made for them. Here are two simple strategies utilising ad serving and simple cookie technology:

1. When an existing customer comes back we know who they are. So in the same way that we would serve up a tailored offer within Internet banking, serve up cross-sell or up-sell offers that are super relevant within the public site or through third parties. Don't get lazy and offer them the same offers the general public is seeing.

2. If a visitor comes to the site and is trawling through the credit card section, personal loans or mortgages landing pages, next time they visit offer to continue the journey. Place a banner on the home page panel that says something like "Still looking for a great mortgage?". This is technically very simple to do; you just need a marketing team that is not thinking campaign, and is instead thinking **offer to target audience.**

Best Practice Website Templates

The other logical question throughout this whole discussion must be what is a "great" Internet presence and how do we achieve it. There are two or three core sites that are a must for every bank, and there is a global best practice that is very easily definable as a result of analytics, metrics, revenue results, SEO penetration, and the like. The three core sites are:

1. public website/homepage (customer led);
2. Internet banking website (customer focused); and
3. portal site (for multinational or multibrand banks).

I'm afraid the investor information, corporate identity site and news/press release websites are absolutely useless on the average homepage unless you are a venture capital or trading company. They don't generate revenue and they are rarely visited (less than 0.5 per cent of visitors to the average retail banking website selects this content), even by the intended audience. For 99.5 per cent of your actual customer base, this content is simply reducing the effectiveness of your site in gaining search engine penetration and revenue leverage. Quite simply, any banking website that still has press releases, corporate information and investor information dominating the homepage are just kissing real revenue goodbye and losing opportunities for cost savings by the truckload. Such content is a waste of your time and your customers' time. In fact, they are costing you significant losses in revenue. Please make the resolution right here and now never to put a press release on your homepage ever again.

> **"Press releases are the bottom feeders of the corporate website ... they offer no value and generate no revenue."**
> Gerry McGovern, author of Killer Web Content

We've already looked at portal sites in the body of the chapter, but I wanted in the conclusion here to give you a simple template of a best-practice site for the public homepage and for Internet banking.

A few rules first. If you have the Public Relations or Corporate Communications team in charge of your website strategy today, fire them. Put in place a Customer Experience team that is focused on customer management and *not telling the customers what you have to say*. Also, I'm afraid that niche areas such as pension funds, wealth management and stock trading need to be relegated to supporting content rather than be dominating the homepage.

Additionally, many within the bank would argue that commercial or corporate banking needs to occupy equal space on the homepage. The fact is, analytics show that this strategy is not logical. Even with the most successful banks that have both a corporate and retail bank, 95 per cent of daily traffic is still retail customers. Thus the homepage should be dominated by retail content and offers.

The public website homepage

So let's look at the public website homepage first. It should be a site that offers the following content and features:

- Access to Internet banking.
- Offers of Internet enabled products (in other words, I can apply completely online).
- A list of the top five to ten problems or issues faced by clients (sourced from the call centre statistics).
- Navigation to all the other areas of the bank.

That's it. Here's what a "wireframe" template might look like for the public website homepage (Figure 5.18). This optimises access to products that actual visitors to the website will actually be interested in and that generate revenue. The APPLY NOW section should be focused on either the offer of the month or the most popular products according to analytics. The marketing campaign panel can rotate with different options, but with no more than four or five different options here, and don't rotate these images too quickly. You could also put the iPhone app download here.

All the products on the homepage should be retail banking products. If other customers, for example corporate or SME customers, are coming to the bank homepage, they simply need to navigate to the second page to see the relevant information and offers. Once again, that page could be

Figure 5.18 Best practice template for a public site homepage
(Credit: UserStrategy, Health Wallace)

similarly constructed to optimise utilisation.

The Internet banking portal

The objective of the Internet banking portal is twofold. Firstly, to give access to all the functions of the bank as respects control over what customers do with their money, and secondly, to target customers with specific offers to cross-sell or up-sell. We have also integrated this with their mobile banking experience, and we are putting our learning from the call centre into active solutions for the customer.

The page is essentially still primarily functional in purpose, that is, to process transactions and save the bank money. However, we've introduced a prominent panel with **targeted, relevant sales offers** for your most valuable website visitors, namely, your existing customers.

When customers come to this site, they want the bank to recommend products FOR THEM. This is not an area for putting the latest promotions or campaigns for all and sundry à la the traditional classifieds approach. If I am a student, I don't want an offer for a Platinum Visa credit card because I know I'll never qualify. If I am a 50-something professional close to retirement age, I don't want an offer of a student loan or a car loan. So here is where the science comes in.

The marketing department is not normally given campaign access to the Internet banking portal, which needs to change as of today. However, more importantly, marketing needs to have the technical and process capability to make relevant offers. This means proper micro segmentation, offer management and linking through to the relevant execution or acquisition technology on the Internet platform.

Business intelligence platforms, such as those of Qlikview™, Cognos and Business Objects, can provide the core information if properly configured for establishing the psychographic or demographic profiles for the right offers. But more than that, there needs to be applications on top of this data mining capability to match relevant offers to the customer. For example, don't offer them an upgrade to a Gold Visa credit card if they already have a Platinum credit card. Let's call this a sales campaign or offer engine. It is the essential piece of kit for Internet banking architecture today, along with a customer analytics engine to feed it with the right inputs or data.

This needs to be largely automated because without it my fear is that the marketing department will simply run out of steam two to three months into this and the same old broadcast, broad-swipe offer approach will dominate these pages.

What's the benefit of this type of selling within the Internet banking secure site? Two primary reasons. Firstly, you get easy cross-sell and up-sell revenue results, and you also have almost zero compliance issues because they are existing customer relationships. Thus, fulfilment is very quick and easy for both the bank and the customer.

Figure 5.19 HSBC Internet banking portal template
(Credit: UserStrategy, Health Wallace)

Internet Channel Improvement Today

So what is on the Internet improvement roadmap that we can achieve in the short term that will bring benefits to both the organisation and to the customer? The following areas represent suggested opportunities for either improvement in financial operations or customer service levels at the branch over the coming three to five years:

- Improved customer content, communications and language.
- Better cross-sell/up-sell capability.
- Better findability and search engine optimisation.
- Improved analytics on customer behaviour.
- Better offer management and generation capability.
- Improved use of application processing automation and service architecture.

These initiatives are designed to optimise your capability to generate revenue and keep customers coming back time and again to interact via the Web. The improvements make themselves evident through a range of projects that can be undertaken within the branch. Some of these projects cross over the above areas of opportunity, so we'd like to list the projects (Table 5.4) as specific illustrations of how improvement and transformation are achievable.

Table 5.4 Achievable outcomes of possible projects

Project/Initiative	Desired Outcome
Usability tests of all sites	Assess any issues with current website language, layout, design and process.
Customer information systems	Improved behavioural analytics on customers across all channels to understand better which "tasks" customers prefer to do in-branch, versus online, etc.
Content management systems	The old dot.com favourite is back, but this time enabled across the organisation so we can "publish" new content continuously. The best analogy is to imagine that your bank is publishing a product catalogue and investor information magazine, based on your product, to customers daily.
Sales intelligence and automated offer capability	Real time and precognitive offer management for existing customers delivered in the form of prompts, offers, or service messages, especially within the Internet banking portal.
BPR (Business Process Re-engineering) on select processes	Reduction of layering between sales and service departments, including the removal of duplicate "skills" within "competing" product units. Creation of "customer dynamics" capability as owners of customers, rather than product competing for revenue from same.
STP (Straight-Thru Processing) and CRM (Credit Risk Management) systems	Enabling customers to get immediate fulfilment for an application rather than waiting the obligatory 24, 48, or 72 hours afterwards due to antiquated manual or human "processes" in the back office. Results in improved service perception and reduction of abandonment due to ongoing process demands (i.e. proof of income, faxing of 3 months' bank statements, salary certificate, etc). Additional benefits include reduction of compliance errors through manual mishandling.
Customer friendly language initiative	Use of ethnography, usability research, audits, customer focused observational field studies and focus groups to improve language and simplicity of application forms and communications with customers within branch (and beyond).
Search engine optimisation	Organic search engine optimisation should be the strategy of every institution, but it requires rethinking what content you actually put on the site because it needs to be driven by what customers are actually looking for.

As discussed in Chapter 1 with the mortgage experience, HSBC Hong Kong had experimented with both lead generation and direct selling online. The limitations of direct selling online were the result of both the interface and automated fulfilment capability; although by 2001, HSBC was already successful selling a limited range of products through the Web, the management took an experimental approach to the Web and allowed the adaptation of a few simple products to the online channel, along with the construction of an online trading platform to support the myriads of day traders in the Asian financial centre.

There had been some limited success early on with the introduction of travel insurance through the public website. By the end of 2001, 2 per cent of all travel and home insurance applications, and 0.5 per cent of accident insurance applications were coming through the online channel. The Personal Financial Services (PFS) management team decided to double the target for online insurance applications in the year 2002, with a target of 4 per cent of all travel and home insurance applications slated for the Web channel.

In reviewing what it would need to improve take-up of general insurance through the Web channel, HSBC looked at how customers were engaged through the website and what information they sought before they committed to the application process.

The site was already getting more than 4,000 visitors a month, so the thinking was to try to capture a few more of these "lookers". They reviewed the site usability, interviewed customers, and prototyped some simple new designs to test on sample groups of potential customers. The results were significant.

What analytics, usability analysis and customer surveys showed was that most of the marketing guff on the existing site, analogous to "brochureware", was simply ignored by customers as irrelevant to the purpose of their visit. Customers simply wanted to know what they would be covered for and what were the conditions in the event of a claim. They also wanted a very simple, easy-to-understand application form.

Users came to the site because they already trusted HSBC; 82 per cent said they wouldn't even bother to compare the price of HSBC's policy with a competitor. Users also knew exactly what they wanted, with 100 per cent saying they were there to apply, not to read, and online was their channel of choice.

The research revealed that while there was already an audience that was willing and able to purchase general insurance online, the site design was complicated and the application process clumsy. In 2001, we required customers to click on five to six hyperlinks before they got to the actual application form. On top of that, we bombarded them with marketing and product speak that was just irrelevant to their core needs when all they wanted was to Apply Now!

The revised site design was simplified dramatically for the general insurance landing page, and in the end HSBC created just five new pages: the "home" or landing page for general insurance with four products, namely Travel, Accident & Health, Home & Motor and Life Protection. The main page offered instant-approval apply-now access to travel, accident and home cover. One-page online application forms were created to simplify the application process dramatically for the customer, along with re-engineering the back-end quotation engine to link through to the website seamlessly.

In fact, the Web changes cost less than HK$80,000 or just a little over US$10,000 in development budget. That was the easy part. Over 50 major business changes had to be made to accommodate the new online process, such as allowing for non-applicant facility (e.g spouse only, child only

Figure 5.20 HSBC homepage, before (top) and after

cover) and payment for non-HSBC credit card holders and simplifying the declarations requirements.

The biggest hurdle to this entire process was convincing the legal and compliance teams of what needed to be simplified in the application process and the subsequent "bank" requirements. The lobbying of the self-service team to the compliance department took over three months of extensive planning meetings and consultations before the new site was "permitted" to be deployed. In examining the process, HSBC learned that many of the things they had asked for were not necessary; they had been added over time to make life easier for the bank, not for its customers.

To illustrate. When taking out travel insurance, the compliance department insists on customers providing their address, proof of their address, along with other personal details proving their identity. However, the insurance product team revealed that such details were only necessary if a customer makes a claim. At the time of the claim, the customer could simply confirm his identity to be eligible for the benefit from the policy.

So what was the result of the revamp?

Website traffic immediately increased, although there was limited marketing of the new site. Between December 2001, when the new site was launched, and Chinese New Year in February of 2002, site traffic increased by 250 per cent. This is largely due to the improved usability of the site. The peaks of activity over the next 12 months correlated heavily with public holidays in Hong Kong.

Initially, HSBC had been receiving about 800 travel insurance applications per month through the branch network prior to the launch of the new travel insurance site. But within a few months of the site being launched, 15 per cent of all travel insurance applications came through the public website. As HSBC supplemented the online activity by offering the same general insurance application forms through the secure Internet banking portal, traffic increased again to 22 per cent of all applications. This growth continued to a staggering 8,000 applications for travel insurance in peak months, and an average of 2,500 to 4,000 in slow months. Currently, more than 78 per cent of all travel insurance applications come through the Web channel for HSBC.

So what happened in the branch? The number of applications through the branch never changed; it remains today at the 2001 levels of around 800 applications per month over the counter. The Web business was simply **new revenue.**

What about the real business metrics? Apart from revenue did this add up for the bank? Over the counter at a HSBC branch in Hong Kong, the average handling cost of a travel insurance application was estimated to be around HK$320–$364 per application. Online, even after fully loading all the development costs and other process changes, the application cost was still estimated at less than HK$85 per person. So the cost savings attributed to the channel were between HK$800,000 and HK$2.3 million per month, just for travel insurance applications. In peak months, new "non-branch" travel insurance revenue totalled more than HK$7 million per month. Not bad for an investment of around US$10,000. Remember too, this was not at the expense of the branch revenue. Branch applications remained the same. This was just new revenue.

The key lesson to be learned here is that the traditional application process did not fit, neither did the traditional marketing and positioning approach. The key was to anticipate customer behaviour, their choice of channel and turn the provision of travel insurance into a service, not a sale.

KEY LESSONS

After 10 years, the Internet is still perceived as a "threat" by some traditional bankers, or at best, just not understood by most bankers.

Far from being simply a "functional" transaction platform to save costs, the Web is the greatest source of new revenue that exists today. Understanding what to sell and how to use the channel in the sale process is the key.

Like for just about every other "alternate'" channel, bankers need to start treating the Web as the equivalent of the branch in strategic importance. Anything less than equal footing simply means loss of new revenue opportunities and loss of customers to alternative providers.

Keywords   Product Mix, Findability, Usability, Value Exchange, Content Scoring, Cross-Sell, Up-Sell, Public Website, Internet Banking, Wireframes, Interaction Design, Best Practice

Endnotes

1   Georgia Tech University Research

2   HSBC Online Internet Survey 2004; Standard Chartered Global Online Survey 2007

3   Various sources: UserStrategy online survey for StandardChartered.com; Alexa.com; GoogleLabs Trends

4   Interview with Countrywide's CTO: *AllBusiness*, 2006. http://www.allbusiness.com/personal-finance/real-estate-mortgage-loans/805072-1.htm

5   GVU's WWW User Survey. http://gvu.cc.gatech.edu/what/websurveys.php

6   NPD Group. www.npdgroup.com

7   Port Washington, N.Y., 19 December 2008. The NPD Group's e-commerce weekly tracking service

8   Microsoft adCenter Internal Data, November 2008

9   Various sources: google.com; alexa.com; NielsenNetRatings.com; SearchEngineWatch.com; etc

10 References: Jakob Nielsen, NielsonNormanGroup, "Usability Testing Sample Sizes", 2004; Jeff Atwood, "Low-Fi Usability Testing", Programming and Human Factors, 31 January 2007

11 See *The Usability Kit* by Daniel Szuc and Gerry Gaffney. Source: http://www.sitepoint.com/kits/usability1/

12 Various sources: alexa.com; www.google.com/trends; UserStrategy surveys for HSBC.com; StandardChartered.com; ADCB.com

13 alexa.com comparison of top 10 banking websites in the US

14 Webtrends data for www.hsbc.com.hk. (93.7 per cent of Chinese-language users and 95.2 per cent of English-language users click on the log-in button for personal Internet banking)

6 Mobile—The New Internet and Death of Cash?

MOBILE banking has always held great promise, particularly in certain geographic locations where demographics and culture have created rapid adoption cycles. From locations such as Sweden, Hong Kong, Singapore and the Philippines where adoption rates reach 70–80 per cent very early in the adoption cycle, the mobile was a phenomenon. Many might argue that mobile banking simply has not lived up to all the hype—and they would be right. Do you remember **WAP?** Well, you can be forgiven you if you don't. Back in 1999–2000 as the Internet dot.com bubble was in full swing, everyone was talking about WAP. Organisations such as Accenture and McKinsey were telling us that sometime between 2004–08, "mobile" Internet usage would exceed the wired Internet in number of connected devices by 10 to 1, and was thus the "place" we had to be digitally. WAP was supposed to be the beginning of the great experiment in mobile commerce. WAP stands for (should I say stood for?) Wireless Access Protocol, but in reality it probably should have stood for Wireless-And-Pathetic.

From WAP to iPhone —
The Emergence of Portable Banking

The fact is that WAP, in its day, was a huge disappointment. There are a number of reasons why it simply didn't work. Firstly, the pre-GPRS wireless Internet was just too slow to be productive as a digital medium. Secondly, handsets did not have the capability or screen size to display information in a meaningful and usable fashion. Lastly, there was no application or content support. Therefore, even if you got past the first two issues you still found menus that were clunky and difficult to use, or you were forced into attempting to use a Web page designed for a normal PC screen on your very slow phone. So the telcos, IT companies and a few bleeding edge bankers got on CNN, CNBC and BBC talking about how WAP was going to make the dot.com bubble look like a bachelor's party for nerds, whereas the mobile paradigm was going to change the lives of everyone from Calcutta to Sao Paolo. And they were wrong ... or were they?

Since the days of WAP, we've had an acceleration of mobile convergence, mobile application development, wireless technologies and other contingent technologies. In the US, while WAP was not even implemented in most networks, the likes of DoCoMo in Japan was trialling 3G networks and new phones, applications and technologies, ranging from streaming audio/video, video telephone calls to mobile shopping.

In the US and UK, while "texting" and BlackBerries were just starting to catch on, the average Philippine mobile user was already doing 188 text messages or SMSes (Short Message Service) per month, or 10–12 messages per individual subscriber per day.[1] Now the US is fast catching up. Today, more than 93 per cent of US adults own a cell phone and the average American teenager sends an incredible 2,272 text messages per month![2]

In recent times, we've had a major leap in hand-held technology with the launch of the Apple iPhone™ and, more recently, Google's Android™[3] enabled devices such as the Nexus One and the Motorola Droid. These new multimedia extravaganzas can only loosely be called mobile phones. While they do allow you to make a telephone call, the fact is that convergence and rich application capability have turned the hand-held "phone" into a major fashion accessory, business tool and productivity device. These days, it's not unusual to find the following standard capabilities in a fairly moderately priced "mobile" unit (Figure 6.1).

Figure 6.1 Standard capabilities in a "mobile" unit

Later in the book, in Chapter 9, we'll talk about Moore's Law. It's based on Intel founder Gordon Moore's prediction back in the 60s in respect of the growth in capability of integrated circuits or microchips which somewhat explains the phenomenal growth of mobile capability from a technology curve. However, to put the above in perspective, think about the number of separate devices you would need to carry around with you 20 years ago just to have the same capability that's available in your mobile today—portable two-way radio or early mobile phone, SLR camera, Sony Walkman, VCR video camera, VCR video player, CRT TV, diary, personal computer (PC) or laptop, generator or car battery, etc, etc. Even then, you still couldn't do half the stuff you do today on that hand-held device.

So we've got all this great technology integrated into these hand-held devices we carry around in our pocket or handbag, and we've got used to them so quickly that they just seem perfectly normal today. This integration of technologies into a single device is what we typify as **convergence** and it is one of the reasons that the mobile Internet is definitely here to stay. The relative ease with which we've adopted such technology into our daily life comes back of course to the **technology adoption diffusion** we discussed in Chapter 2.

However, convergence makes such growth in the mobile arena even more compelling as an argument for participation. Convergence drives adoption rates even harder for consumers when it comes to mobile. Just think about how many of your friends or colleagues don't actually own a mobile phone.

Mobile is here to stay and it's going to make an even bigger impact in the way we bank, if we can just get the delivery platform right. To understand what elements of the mobile phenomenon make for important building blocks for retail bankers, we're going to look at the following key areas:

1. **Device hardware.** We've had Motorola, Nokia, Apple, SonyEricsson, Palm, BlackBerry and others. But where is the mobile device going, and what does it mean for banks?

2. **Operating systems and software stacks.** There have been various players in this arena, but the most notable are Nokia Symbian, Google Android, Apple OSX, RIM's BlackBerry

OS and Palm WebOS. To simplify this discussion, we'll call these operating systems "software stacks", that is, platforms to enable application development.

3. **Needs and solutions.** What role does the mobile play in the future? There are three key solution areas, namely mobile banking, mobile payments and a vehicle for assisting the great unbanked.

4. **Interface.** Unlike your laptop or desktop, the screens on your mobile device are relatively small, so how does this affect your ability to deliver content and functionality?

5. **Bank integration.** We'll look at the supporting technologies within the bank and through third parties that make mobile banking viable

6. **Revenue opportunities.** SMTM or Show Me The Money!

| Device Hardware | OS or Software Stacks | Needs and Solutions | Interface | Bank Integration | Revenue |

Figure 6.2 **Key areas to consider in mobile banking development**

Ultimately the objective of this chapter is to demonstrate the significant changes in mobile usage historically, why the mobile device will continue to dominate our daily lives, and how this has an impact on banking. Many bankers today are rightly sceptical about the real impact of mobiles beyond SMS alerts and some simple transactions. However, the growing capability of these devices opens up some very exciting opportunities—and the great thing is that the technology is available to try much of this today.

While available almost anywhere on the planet, the growing trend is for mobile devices to get smarter, more capable, and be much more of an everyday platform for the average user. Even my nine-year-old daughter who is demanding her own mobile will tell you the same.

Device Platform—Why Motorola Failed, and Why BlackBerry Still Might

At the time of publication, Motorola's mobile division is under significant operating pressure, and some analysts expect them to announce the closure of this business unit shortly.[4] Despite having a long and glorious history as one of the staple brands in the mobile arena, this long held brand is under significant pressure because they are focused too much on the engineering of the product and not enough on changing customer behaviour.

In 1983, Motorola, who already had a very successful car phone division, produced the DynaTAC8000x, the first commercially available cellular phone small enough to be easily carried around, but it retailed for almost US$4,000 ($8,544 in present-day terms.) It wasn't until the mid-90s that such mobiles started to become mass market devices carried by more than the elite early adopters. This early innovation led Motorola to dominate the early mobile phone market and they stayed at the number one slot for more than 15 years. Motorola's amazing history as a producer of professional radio gear served them well in the shift to the cellular market. In fact, it was a Motorola Radio transponder[5] that transmitted Neil Armstrong's first words from the moon back to planet Earth. Their position was so strong that it wasn't until 1998 that Nokia was in a position to displace Motorola as the number one producer of mobile. Unfortunately, Motorola's woes have steadily continued.

Motorola built its mobile division on technology innovation. When it launched, there was no one else even close to developing a competitive offering, and indeed it took years for the competition to catch up. The very first "clamshell" mobile phone, the Motorola StarTAC, was considered by some to be one of the greatest gadgets in the last 50 years.[6] Despite this phenomenal market leadership, the shine was starting to come off by 1998.

Motorola had always had central to its commercial platform the capabilities of the handset from a functional perspective—the core technology and functionality. This comes from the strong engineering pedigree of its mobile phone division. This functional-led approach enabled Motorola to excel with the smallest and most capable platform in the 90s. But by the mid-90s, Nokia was hammering at Motorola's door, particularly

Figure 6.3 The Motorola StarTAC, circa 1996 (Credit: Motorola, Wikipedia)

with the launch in 1996 of the 8110 (sometimes referred to as the "banana phone") and then shortly thereafter with the 9000 Communicator. Motorola had only a few variations in its range at the time, so Nokia's approach to try to attract different segments and demographics worked well. But Nokia also focused on innovating around the software interface. Simply put, Nokia's phones were easier to use, which appealed to a wider audience and attracted a whole new slew of customers.

The RAZR mobile, with big budget and successful advertising, saw a brief resurgence in Motorola's success in the mid-2000s, but the success of the BlackBerry in the corporate arena, new handsets from LG, Samsung, Sony Ericsson and the iPhone's launch left the RAZR wanting. Primarily this was due to limited application, Internet and browser support—effectively what amounted to an ageing OS and firmware approach. In its latest move, Motorola has announced its pairing with Google Android and its Droid app phone in a hope to stave off the complete collapse of their mobile division. It is too early to tell whether this move will be successful.

The lesson, however, is clear. A shift in user preferences occurred in the mid-90s to form over function, devastating Motorola, which did not adapt from purely function-led hardware. Motorola found a brief resurgence in brand popularity with the RAZR, pushing the phone as a fashion accessory, but that too was short lived. The shift in user preferences has again morphed with an emphasis on device usability and connectivity, largely as a result of the iPhone's popularity and more capable devices with MP3 and wireless capability.

So what will come next that will drive the platform and success of smart devices or app phones? There are a few drivers morphing the requirements of hardware and device platforms moving forward. The dominant drivers will be the increased utilisation of applications, wireless content delivery and data services, increased convergence, clustered social behaviours and new materials science.

In the next 10 years, integration of devices into wearable computing platforms will be popular, as will more "fringe" designs based on experimentation with more exotic materials. Delivery of personalised data services and content will be key as we shift increasingly to subscription-based revenues for application providers and network operators. For financial institutions this will have some fundamental and earth shattering implications.

Future phones are going to be all about the **how**—that is, how we use the device, how the device fits into our lives and social context, and what connectivity it provides to the services we use. Device convergence will continue with our devices increasingly communicating with the outside world, monitoring our health and well-being, searching out our friends and "tribes" in the nearby ether, and emoting more about our current social status or interactions.

From a banking perspective, the phone will connect us seamlessly to investment, payment and merchant opportunities. These devices will talk more frequently to the bank, our telco, airline loyalty programmes, retailers and such to provide us with an optimal consumer experience based on our connections. No more having to log in and check your miles balance; this will be integrated into an app on our smart device.

Likewise the devices themselves will get more integrated into our lives.

Figure 6.4 The Nokia Morph and Aeon Concepts (Credit: Nokia)

There is a practical limit to how small phones can be because of battery and screen size, but just as you might choose a different style of watch or item of jewellery for different occasions, you might do the same with your phone. Additionally, the limitations in respect of how small phone hardware might be are fast disappearing, as can be seen by the Caps concept phone (Figure 6.5).

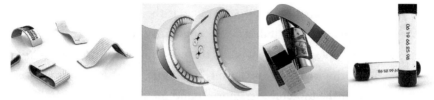

**Figure 6.5 (from left) Nokia 888 (Tamer Nakisci), Benq-Siemens "Snaked",
NEC and "Caps" concept phones**[7] (Credits: Tamer Nakisci, Benq, NEC)

These smart devices will have to work seamlessly with our environment, and they will need to be increasingly versatile. As the technology gets smaller, manufacturers and corporations will get smarter about how we can utilise it effectively. As wireless access gets faster, advertisers can push more information to the invisible data pipe, so they will have to be very careful not to "burn'" customers with irrelevant information. Then again, it won't be long before we'll have spam filters trapping irrelevant offers on our mobile phones just as we do with email.

> "The computer in your cell phone today is a million times cheaper and a thousand times more powerful and about a hundred thousand times smaller than the one computer at MIT in 1965, and so that's a billion-fold increase in capability per dollar or per euro that we've actually seen in the last 40 years.... The rate is actually speeding up a little bit, so we will see another billion-fold increase in the next 25 years and another hundred-thousand-fold shrinking. So what used to fit in a building now fits in your pocket, what fits in your pocket now will fit inside a blood cell in 25 years."
>
> Ray Kurzweil, Inventor and Futurist[8]

Operating Systems and Software Stacks— Driving Mobile Application Development

Many phone manufacturers have clung to their own device paradigms because of management ideals and specific design processes. Manufacturers such as Motorola (until very recently), Palm and BlackBerry have hung on to their proprietary operating systems and core hardware strategy. This in itself is not necessarily a problem, except supporting content and

application infrastructure for devices such as the iPhone (e.g. iTunes[TM9])
has made application capability a key competitive edge.

Despite Apple's success, all is not 100 per cent rosy on the iTunes front
either. With the average expectation for iPhone apps being either free,
99 cents, or at the top-end averaging around $3.99 per download, iPhone
developers are blogging about the lack of real revenue being achieved through
the platform.[10] Apple has also judiciously avoided implementing Adobe
Flash[TM] in its iPhone Safari browser because it would mean developers
would no longer have to work with iTunes. All this aside, the facts are
users want content and rich functionality on the devices they choose, so
there needs to be the ability to leverage the communities of developers to
generate these.

To be competitive, manufacturers will increasingly have to offer
a wide range of applications for their devices, and it makes sense that
device manufacturers will not have the resources to develop all of these
applications in-house. Thus, communities of mobile app developers are
now a much sought after resource.

While manufacturers may seek to retain their proprietary systems,
unless they have a strong third-party strategy to encourage data services and
application development, these OS dependencies will definitely reduce their
competitiveness if they are not "open" enough to developer participation.
The Google Android strategy of an open platform approach is a very
attractive strategy for developers. Nokia's acquisition of the QT platform
also leads along the lines of this strategy. The key issue still facing developers
is lack of consistency from one mobile screen to another, and in respect of
input methods (i.e RIM keyboard, multitouch, stylus, keypad, etc).

In June 2008, Nokia acquired the remaining share of Symbian, the
company which developed the open source mobile operating system of
the same name. Symbian is currently deployed on Nokia, Sony Ericsson,
Panasonic, Siemens and Samsung's platforms. However, anticipating the
increasing demand for application content, along with the success of iPhone
and the buzz around Android, Nokia has also acquired Linux developer
Trolltech®. Trolltech built QT, a software stack or toolbox for developers,
in January 2008 for $153 million. QT is a cross platform, open source

development toolkit used to develop such applications as Skype, Google Earth, Opera and Adobe's Photoshop Album.

RIM/BlackBerry (NASDAQ:RIMM), on the other hand, is relying on its penetration of the enterprise/corporate solutions market. Last year, RIM/BlackBerry had a 16 per cent market share in smartphones globally, and they have increased that to 20 per cent in the first half of 2009. Pretty impressive in a market where total mobile handset sales declined dramatically. The Apple iPhone does not even register as a competitor to RIM in the enterprise/corporate market. Why does BlackBerry have such a strong hold? **Wireless access to corporate email and a device that is purpose built for mobile responsiveness to such demands.** That's it, folks—it's not application platform or hardware functionality. The BlackBerry is simply the easiest and most stable platform for the corporate market because its support for corporate email is rock solid. The lack of an application platform, however, could be BlackBerry's downfall if they aren't careful.

So why can't the iPhone, which has taken the world by storm, compete with the BlackBerry in this space? The key problem is its onscreen Qwerty keyboard mechanism. While the iPhone looks cool and the removal of a physical keyboard reduces the hardware footprint and increases usable screen size, the onscreen keypad is simply not as accurate as the RIM/BlackBerry physical keyboard.

Studies show that the virtual keyboard on the iPhone allows users to enter text as rapidly as with a RIM physical thumb keypad, but the Apple virtual keyboard simply produces greater inaccuracies. In a test of the comparative methods, BlackBerry users thumbed out communiqués as quickly as iPhone users did. The iPhone users, however, made an average of 5.6 text-entry errors per message, compared with the BlackBerry set with just 2.1 errors per message.[11]

Figure 6.6 The iPhone on-screen keypad

The key to the future of mobile banking is not a specific device, but it is clearly the ability to generate applications for mobile smart devices and not just rely on limited SMS or mini-browser capabilities. Here are a few examples of the beginnings of such forays into mobile bank applications.

Westpac, BofA and Chase: iPhone app

Westpac, BofA, Chase's iPhone apps integrate with your Internet banking account to pay bills and credit cards, see your account balance and mini statements, transfer money between your accounts. They also allow you to use the iPhone's built-in GPS and Google Maps capability to locate the nearest branch or ATM.

BBVA Compass: iPhone app

BBVA Compass is a bank with a presence in more than 30 countries. Their iPhone app offers a more user-friendly experience. The phone gets synced with your bank account so you don't have to log in every time to get your balance. And BBVA only sends the last 4 digits of your account, so risk is minimal even if you lose the phone.

Lemonway: Wonderbank app

Lemonway's cross-platform Wonderbank app provides secured log-in, balance enquiries, account details, funds transfer, bill payment, fraud alert, stock quotes, portfolio management and other possibilities. They also have an advanced version designed for traders on the move called Wonderbank+.

In 2009, BlackBerry, Nokia, Samsung and others launched app stores to compete with Apple's iTunes success. With Apple already carrying more than 100,000 apps and 1.5 billion app downloads in the last 12 months, it's difficult to see how other platforms will quickly compete.[12] Apple's iTunes success was phenomenal as they only started in 2007 when the iPhone was launched. Thus, others could gain ground just as quickly.

Figure 6.7 The iTunes apps store

Coming from a different strategic direction, however, Google does not see a head to head conflict with iTunes in the long term. Given they have launched a capable iTunes competitor, Android Market™, but Google's plans for mobile application stack domination comes in the form of cloud-based, Web-centric apps that exist on a central server. A central server allows users to access apps on any mobile platform as long as it has a mini-browser. Google believes that this approach will win out over iTunes, which is inescapably linked to the iPhone platform. A whole underground apps market exists for the iPhone based on a method called "jailbreaking". This shows that while iTunes has been phenomenally successful, not every user is happy to have this as the sole solution to their content needs.

Google will be relying somewhat on the next generation of Web browser programming code, namely HTML5, which is in the pipeline[13] as the next standard for the Web hypertext markup language. It is expected to be a game-changer in Web application development, making obsolete plug-in-based rich Internet application (RIA) technologies such as Adobe Flash, Microsoft Silverlight and Sun JavaFX. Thus browser interfaces will act much more like software applications than Web pages. For now we call such websites for mobiles, mobile portals, or **m-portals.**

HTML5, if it lives up to its promise, could very well be a revolution in mobile browser capability. But by definition, it will require more bandwidth to be viable because the markup code will be more complex and heavier. Using fully-featured HTML5 on today's browsers and mobiles might be likened to trying to download YouTube videos or flash websites on a dial-up connection. But with 4G wireless, WiMax technologies and faster processers, HTML5 will provide a very rich tapestry for future smart devices. In fact, within the next 10 years we will be looking at devices capable of 1 Gbit downlink speeds.[14] To put that in perspective, in the future your mobile will be able to download a DVD quality movie from anywhere in about half a second.

The key here is to understand that users fundamentally want **rapidly delivered content** and **functionality** in a user-friendly environment. What banks need to do is figure out what the killer apps for mobile are. In the following sections we'll try to give you a few ideas what these might be.

Needs and Solutions—Why Mobile is a Viable Business

To break it down, let's analyse the various pieces of data available for information or services that customers might want access to when they are on the move.

There are really two broad categories of solutions required here. The first is **content as a service;** the second is access to key **functionality or financial services platform capability.** These are articulated in three major solutions areas offered through mobile enablement. They are:

1. **Mobile bank.** This is the bank-in-a-pocket concept.
   What is possible, and is it financially viable for the bank?
   We'll also look at some case studies of how to do this right.

2. **Mobile payments.** We look at how mobiles are being used to revolutionise payments, especially in the micropayments arena.

3. **Banking for the great unbanked.** For the millions of individuals out there who don't have a bank account, are mobiles the solution?

Let's review the first of these, the so-called "mobile bank", or bank-in-a-pocket concept.

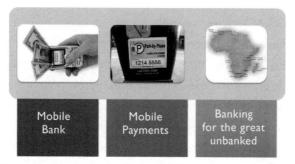

Figure 6.8 **Key solution areas through mobile enablement**

The mobile bank

Access to banking services is right up there as an everyday basic functional requirement. If you look at bank call centre traffic in the US, Europe, Asia and the Middle-East, the consistent number one call type or request either via IVR or through the contact centre is Account Balance Enquiry. So it would be a no-brainer to be able to provide this in a secure way through a mobile device at the start of each day to the customer, or on demand as required.

Don't let the IT guys tell you that this is technically difficult, or that it is difficult to guarantee security. Most mobile phone mini-browsers and application platforms provide 128-bit SSL encryption as a minimum, the same essential security layer we have behind secure Internet banking. So what's the problem? The issue generally is the bank's willingness to do this or an internal management culture that underestimates the mobile channel, not the bank's ability to execute. How do we know? Well, the fact that Chase, BBVA, Westpac and others are already doing it is a good clue.

We already deliver the account balance to our call centre systems, and you can access your account balance via ATMs or Internet banking, so what's the big hang-up about making this available through a mobile app or m-portal? Secure SMS and MMS technology has been available since 2005. Additionally, we could embed encryption technology either in an application, through a secure session or even via SIM-based security for added security.

What would be the effect of giving customers daily access to a simple account balance subscription or broadcast service? Firstly, it would reduce call centre and ATM activity. Secondly, by introducing payment reminders and other such additional services to this facility, it would improve the involvement of the customer day to day with the bank. If your bank doesn't take this step, then it won't be long before third parties step into the breach to take up the opportunity.

MoBank of the UK is a good case study in this respect. MoBank currently supports **balance look-up** with a gaggle of banks such as the likes of Abbey, Bank of Ireland, Halifax, Bank of Scotland, Cahoot, Co-operative Bank, First Direct, HSBC, Intelligent Finance, Lloyds TSB (UK), Nationwide, NatWest (UK), RBS and others. MoBank also offers a range of retail products and services through the iPhone app. It works with your existing bank account, credit or debit card, to let you buy and pay for items using your mobile phone. MoBank users can even browse for flights, pizza, flowers, gifts, or any other goods or services, and pay for these securely.

MoBank has realised what few banks have—that customers are not only ready for mobile banking applications, but they are ready for

Figure 6.9 MoBank offers a mobile bank aggregation platform

more. The bank as a "utility" is all well and good, but unless you can utilise the bank's utility for the sort of payments and activities you are interested in as a consumer, you'll go looking for alternative solutions from third parties. Don't believe me—then what are you doing with a credit card?

With over 100,000 apps on the iPhone platform, it is clear Apple is doing something right. You might assume that the top applications sold on the iPhone are games or media content, but this is not the case. As you can see from the list of Top iPhone Apps of 2008[15] (Table 6.1), most could be classified as productivity tools—basically tools to get you through your day or help you utilise the unique properties of the device, such as the camera, GPS or MP3. What this illustrates is that consumers are not just looking for static content, games and entertainment, they are looking for solutions to day-to-day problems.

Table 6.1 Top iPhone Apps of 2008 (Source: Time.com)

App	Description
Pandora	Free songs streamed weekly to your iPhone
Midomi	Ingenious song selector, hum a tune and it finds it on iTunes
Yelp	Food guide and review based on user ranking
ShoZu	Upload photos to various social networking sites
NY Times	RSS style access to NY Times news as it breaks
WritingPad	Simple document pad that uses a clever method of entry
Stanza	Access to over 20,000 free classics in the public domain
Jott	Transcribes 15-second voice memos into text
Google Mobile App	Finds anything you need based on Web search and phonebook look-up
Instapaper	Uploads articles you find online from your browser to your iPhone

According to recent research, the average Brit is more likely to divorce than switch banks.[16] So perhaps this produces complacency in respect of the banking relationship. What mobile banking does, however, is allow for a much higher degree of customer satisfaction and keeps the brand at the top of the mind, allowing for future cross-sell/up-sell opportunities, all at a fraction of the cost of traditional service delivery mechanisms. All it requires is thinking about a different interface strategy and what can practically be executed on a mobile device.

If you have any reservations about whether this is good banking, just look at the take-up of Bank of America's mobile banking iPhone app. They launched in August 2007 and had 500,000 customers by November 2007, a million by June 2008 and 2.2 million by April 2009. Today, they have more than 3.5 million mobile customers. Let me assure you this is not because BofA is a great bank—it's because customers want mobile.

Translating ATM, call centre and Web analytics into a Web app

Table 6.2 overeaf is a snapshot globally of the top five monthly requests or active transaction demand via ATM, call centre and Internet banking. This is taken from a series of research projects, analytics, customer focus groups and surveys.[17] But from market to market, the trends are generally consistent, with some local variations around specific bill payment types or something similar. The data can easily be verified for your institution by checking for transaction usage data on the ATM, call centre and IVR transaction statistics, and total Web analytics (probably through Web trends or similar).

Table 6.2 Top five transactions across channels

ATM	CALL CENTRE	WEB CHANNEL
Cash withdrawal	Account balance	Account balance
Account balance	Credit card balance	Bill payment
Transfers	Recent transactions	Transfers
Deposit	Bill payment	Credit card balance
Bill payment	Lost credit card	Pay credit card

So the above is actually the perfect base functionality for an iPhone or Android application for ALL your customers. The only one of these we can't do is the cash withdrawal function, although mobile payments will make this increasingly superseded.

There are really two classifications of bank-enabled functionality embedded here in an application. The first classification is transactional, the second is content. Account balance, credit card balance, recent transactions, loyalty programmes, miles/points balance, and bill payment content can be data that is streamed to a customer's phone. In fact, customers may even pay for this as a service. Functionality such as transfers, initiating a bill or credit card payment needs interaction through a transactional platform.

The clear proposition here is that the customer registers his application through the bank or through the app store and puts in an initial level of authentication into the application so that the streaming data can be delivered or refreshed each time the app fires up. If the account information shown is limited to the last four digits of the account number, there is hardly a risk of abuse. Even if the phone is lost, the streamed data is just informational with no content that is open to abuse through fraud.

The additional transactions such as transfers could be confirmed with a secure login or authentication each time, just like with Internet banking, but with the added possibility of SIM card-based encryption thrown in.

Mobile payments—the Holy Grail of mobile transactions

To say mobile payments are inevitable would be an understatement of the highest order. We could devote an entire chapter or more to this subject alone. The facts are, for both current mobile users and the great unbanked, mobile payments are a solution that begs to be implemented and are already taking the world by storm.

Mobile payments refer to the grouping of alternative payment methods. Increasingly widespread in Asia and Europe, these alternative methods are particularly effective for micropayments. Instead of paying with cash, cheques or credit cards, in a mobile payment scenario a consumer can use a mobile phone to pay for services or goods. Goods and services that can already be bought with a mobile include music, videos, ringtones, online games, wallpapers and digital goods, transportation fare (bus, subway or train), books, parking fees, tickets, fast food and other hard goods.

There are four primary models for mobile payments:
- Premium SMS-based transactional payments.
- Direct mobile billing.
- Mobile Web payments.
- Contactless payments (NFC, RF-SIM).

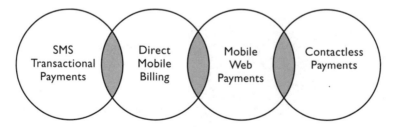

We started playing with mobile payments services way back in the 90s (I'm feeling old saying that right now, but it was 20 years ago). Sonera, a mobile operator and services provider based in Finland, started to offer mobile payments back in 1997. The first payment options included the ability to purchase soft drinks from vending machines. In 1999, Payway piloted parking meter payment in Stockholm. Sonera (now TeliaSonera) then launched Sonera Shopper in 2002, featuring WAP enabled payments on a range of products and services.

The combined market for mobile person-to-person (P2P) payments is expected to reach more than $60 billion globally by 2013,[18] while the mobile payment market for goods and services, excluding contactless transactions and money transfers, is expected to exceed $300 billion globally by 2013.[19]

In Asia, mobile payments have taken off like a storm. Last year, four million South Koreans bought music, videos, ringtones, online game subscriptions and articles from newspaper archives and other online items, and charged them to their mobile phone bills—every month—without going through their bank or using their credit card. This amounts to total mobile transaction revenues of 1.7 trillion won, or approximately US$1.4 billion, in 2008 alone. **T-Money**™—electronic cash stored and refilled in SIM cards and phone chips—can be used to ride the subway and bus or buy snacks from a 7-Eleven store, vending machines or cafeterias at school. Instead of giving their children cash, Korean parents now transfer money to their kids' T-Money account.

T-Money also makes mobile gift giving possible. Someone can check into a mobile carrier's online shop, buy an icon depicting a Starbucks frappuccino and send it to his girlfriend's phone. She can then go to the Starbucks, show the icon and get the drink. Each day, 70,000 mobile gifts, from Dunkin' Donuts and pizza to underwear and cosmetics, are delivered via SK Telecom's networks.

Figure 6.10 T-Money contactless application for public transport and payments (Credit: *International Herald Tribune*)

In September and October of 2009, both Starbucks and Amazon launched mobile payment solutions for their US customers. Many Starbucks customers will already be familiar with the Starbucks Gift Card, a stored value card which can be used as a replacement for cash to make purchases at Starbucks. The Starbucks iPhone app replicates the Starbucks card, allowing users to pay for their orders at point-of-sale with

Figure 6.11 Starbucks virtual payment card via the iPhone (Credit: Starbucks)

their iPhones. Starbucks has installed special barcode scanners that are capable of scanning QR code tags or semacodes, a type of bar code that is readable from a mobile screen.

E-money and mobile payments started in Japan in 1999 and usage is growing exponentially. They are today are an important and big part of Japan's economy. Seventy-two per cent of Japan's population are active and paying mobile Internet users, which makes it an excellent platform for the expansion of mobile cash. In 2003, SONY's FeLiCa IC semiconductor chips were combined with mobile phones to introduce the first "wallet phones" ("Osaifu keitai" – おサイフケータイ). Today, the majority of mobile phones in Japan are wallet phones.

The two parallel systems in Japan today are **Edy** and **Mobile Suica.** Edy stands for Euro, Dollar, Yen—expressing the hope for global success. Intel Capital believes in this success and has invested in the company that runs Edy—BitWallet (backed by SONY). Mobile Suica is a service for Osaifu Keitai mobile phones, first launched on 28 January 2006 by NTTDoCoMo and also offered by SoftBankMobile and Willcom. Suica is a prepaid rechargeable contactless smart card mainly used to pay for fares on the JR East railway network.

By the end of 2010, it is estimated that more than US$3.6 billion or ¥327 billion of transactions will be made by Suica-enabled mobile phones. Edy will account for at least another US$1.4 billion in trade by that time. According to the Bank of Japan, in 2009 mobile money accounted for

Figure 6.11 (left) FeLiCa technology underlies NFC applications with SONY/DoCoMo solutions

Figure 6.12 NFC mobile for transport use (Credit: Wikipedia, Creative Commons)

more than 2.3 per cent of all banknotes/cash in Japan, and within four years we expect to see more than one billion mobile e-money transactions per month.

AmazonPayments™, a type of digital currency and payment platform, has put Amazon into the online and mobile payments fray with their recent iPhone app launch. Like PayPal®, Alibaba's AliPay, Tencent's QQ coins, Second Life's Linden dollars, all these virtual currency players are trying to export their currency to the mobile platform as a means of transferring money or buying goods, services and gifts securely online, or on the go.

Banking on the great unbanked

Although mobile payments are gaining traction in more developed markets, peer-to-peer m-payments such as mobile money transfers are an established and fast-growing fact of life in many developing economies. A large proportion of households in developing countries lack access to financial services, which impedes economic growth and development. A large body of evidence shows that access to financial services, and indeed overall financial development, are crucial to economic growth and poverty reduction.

A lack of formal financial services infrastructure and activity limits market exchanges, increases risk and affects opportunities to save. Without formal financial services, households rely on informal services that are associated with high transaction costs. Thus, increasing access to formal financial services to the majority of households in developing countries remains an important policy goal for institutions such as the UN, World Bank and IMF. It has also been recognised that even for those with bank accounts, physical distances to branch banks or points of financial service add significantly to transaction costs.

"If I leave my wallet at home, I may not notice it for the whole day. But if I lose my cellphone, my life will start stumbling right there in the subway." 21-year-old Kim Hee-young, Sookmyung Women's University, NY Times article, May 2009[20]

Mainstream financial institutions generally shy away from developing economies because of the premise that low-income populations do not save and are bad borrowers. However, the microfinance revolution effectively shattered these myths by demonstrating that when poor households have access to financial services, not only do they save, but they also have high repayment rates and low default rates when they borrow. Muhammad Yunus, the founder of Grameen Bank (essentially meaning "Village" Bank) in Bangladesh, was awarded the Nobel Peace Prize in 2006 for his efforts to revolutionise microcredit on the subcontinent.

Beyond microfinance, however, one of the largest sources of income for developing economies these days is the large population of expatriates living and working overseas, who remit funds back to their families in their home countries. Peer-to-peer money remittances enable an expatriated worker to send money across international borders to family or friends.

According to the World Bank, 175 million migrant workers each year send billions of dollars' worth of international remittances to family and friends, many of whom do not have bank accounts. In 2008, recorded remittances to developing countries exceeded $328 billion, up from $221 billion in 2006 and almost three times the level reached in 2002.[21] India, Mexico, China, the Philippines and Poland were the top five recipients of remittances in 2008, accounting for more than one-third of all remittances received by developing countries.

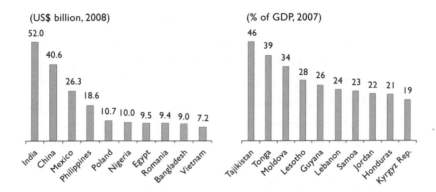

Figure 6.13 Top recipients of migrant remittances among developing countries in 2008 (Source: World Bank)

Nearly five billion people worldwide have little or no access to traditional financial services due to a lack of ATMs and bank branches, poor regulation, low levels of financial literacy or other weaknesses in infrastructure. Clearly, with the wide reach of mobile phones which now outnumber ATMs by 2,000 to one, mobile operators have a potential solution to the access problem and can extend these remittance services to millions of people in remote, rural areas with a relatively inexpensive alternative to expensive private money transfer services.

With the help of the mobile phone, the GSM Association[22] (GSM refers to Global Systems for Mobile Communications, the primary standard for digital mobile phones in use by 80 per cent of the global mobile market) estimates that the international remittance market will grow to $1 trillion by 2012. ABI Research, meanwhile, predicts that the global mobile fund transfer market will generate $8 billion in revenue for mobile operators by 2012—from just over $10 million in 2006. Edgar Dunn[23] (Mobile Banking and Payments Consultancy) research estimates that by 2015, more than 1.4 billion people will be utilising mobile payments services.

A case in point are the mobile telephone money transfer services that allow mobile phone users to make financial transactions or transfers across the country conveniently and at low cost. The two most successful of these are M-PESA in Kenya and G-Cash in the Philippines. (See case study on M-PESA on the following pages.)

M-PESA Success Story
Mobile remittances taking
the developing world by storm

Kenya's mobile payment service, M-PESA (provided by main mobile phone company, Safaricom, in conjunction with Vodafone), represents a good example of how low-cost approaches using modern technology can effectively expand the financial services frontier. M-PESA (M for mobile, PESA is Swahili for "money") is the product name of a mobile-phone-based money transfer service that was developed by Sagentia (now owned by IBM) for Vodafone.

The concept of M-PESA was initially to offer microfinance borrowers a convenient way to receive and repay loans using Safaricom's network and their air time resellers. As there is a reduced cost of dealing in cash, microfinance institutions (MFIs) can offer more competitive loan rates to their users who would gain through easier tracking of their finances.

In 2007, over 70 per cent of Kenyan households did not have bank accounts or relied on informal sources of finance. When the service was trialled, customers adopted it for a variety of uses, but complications arose with Faulu, the partnering MFI. M-PESA was refocused and launched with a different value proposition—sending remittances home across the country and making payments. Today, millions of Kenyans use M-PESA to make payments, send remittances and store funds for short periods. Many of those without bank accounts are able to use this service at low risk and cost. As noted in a recent article in *The Economist*,[24] Kenya's M-PESA is probably the most celebrated success story of mobile banking in a developing country. What started as a mobile money transfer service has become a success story of financial services development with a technological platform that makes it cost effective and safe.

In July 2007, there were 268,499 registered M-PESA customers and by September 2009, the number of registered customers was 8.3 million—an increase of over 3,000 per cent. This is about 22 per cent of the population of 35 million covered in a two-year period. Also impressive has been the increase in the number of monthly transactions, which increased by 4,627 per cent over the same period (354,298 in July 2007 to some 16.7 million in July 2009). In July 2007, the total value of monthly transactions (deposits and withdrawals) was 1.065 billion Kenya shillings (US$14.2 million). This figure was Ksh. 40.176 billion (US$535.6 million) in July 2009, a growth rate of 3,671 per cent. These numbers show impressive growth in the utilisation of mobile payments within a relatively short period of time.

With over 14,000 outlets and reseller agencies around Kenya, M-PESA outstrips the top four banks' reach by more than ten to one. Which is why M-PESA has become ubiquitous so quickly.

M-PESA has now expanded its field abroad. In October 2009, Safaricom launched its M-PESA services in the UK through Western Union, Provident Capital Transfers, KenTV and others. While there are some AML restrictions on the usage of M-PESA for transfers by a single

individual, the system still allows a Kenyan working in the UK to deposit Pounds or Euros in the UK with a remittance agent, and have his family or associates collect that money in Kenyan shillings back in the home country with the use of their mobile phone.

M-PESA has extended its reach further across Africa with

Figure 6.14 M-PESA outlets

its launch of M-PESA in Tanzania. While the take-up in Tanzania has been slower, there are still more than 300,000 users in that country just a year after its roll-out. Vodafone has also partnered Roshan to provide M-Paisa, a local variant of M-PESA, in Afghanistan. Plans to expand the service across India, Egypt and South Africa are also currently underway.

The Edgar Dunn research also found that the number one barrier to successful deployment of mobile payments and wallets was government regulation. Mobile operators and collectives such as the GSM Association are lobbying governments to ensure that regulation governing the deployment and usage of mobile financial services is proportionate to the risks involved.

As reported in the *Nairobi Star* in December 2008,[25] M-PESA with all its success represented a clear threat to the Big Four banks in Kenya, which have combined market coverage of three million bank accounts and 750 banking outlets. M-PESA, in comparison, has more than 8.5 million customers and 14,000 sales agents and outlets across the country. The massive threat that M-PESA holds for the Big Four banks is patently obvious. A similar story can be told in other markets where new payment mechanisms have been successful. The problem for mobile payments in this environment is to what extent do such mechanisms impinge on banks, and should they be regulated as banks are?

Regulators and governments probably do need to provide an infrastructure for mobile payments to be truly successful. This framework might include:

- **Regulation of low-risk money transfer services.** This would involve small amounts of money compared with traditional banking services, outside traditional banking regulation.
- **Enabling non-bank organisations to facilitate the transaction.** Either as an agent of a bank or a remittance provider to facilitate the cash in and cash out activities on both sides of the mobile money transfer.
- Whenever possible, **implementing regulations on the systems** level without interfering with the customer interface.

What is the outcome for banks? Well, as G-Cash from Globe in the Philippines, and M-PESA from Safaricom in Kenya show, the biggest threat to the banks is from telecom operators. Banks need to team quickly with network operators so as not to find themselves competing against these. Given the limited number of network operators in each market, banks should move quickly in case they get locked out by exclusivity agreements or other considerations. To illustrate, the Bank of the Philippine Islands (BPI) and Globe Telecom have recently announced the launch of a mobile microfinance institution, PSBI (Pilipinas Savings Bank).[26] PSBI is a traditional bank that has been converted for use in the mobile and microcredit arena.

Secondly, rather than treat mobile payments as a threat, banks need to see it as an opportunity to open otherwise unprofitable markets for low-income segments. Banks will need strong partners and a strong platform to succeed.

Interface—Small Screen, Big Possibilities

Mobile screens are generally not very large. From the likes of the Nokia 6010 with a screen resolution of 96×65 pixels, to the iPhone with a whopping 480×320 pixels, mobiles are still far less than the old 640×480 Windows 3.11 standard of the mid-90s. While resolutions and screen quality will continue to improve, it is impractical to carry around much larger devices than we already carry around today in our pockets. So either we try to adapt our content to smaller screens now or wait till we are all carrying around hi-resolution, nanotech soft screens or electronic paper in our pockets.[27]

One of the reasons mobile banking has not really taken off as yet is that banks assume that customers can access the current websites and Internet banking interface through the mobile phone mini-browser. But to put that idea into context, the average banking homepage these days is 1024×768 resolution or higher, so even on the biggest mobile screens, you still only get roughly a third of the screen size. So you either see, say, 10–30 per cent of the available content, or you have to shrink the screen to a point where it is no longer readable or usable.

So, the key for banks today is really to have a **separate interface design strategy** for mobile content. This needs to start happening now, because within the next five to ten years we'll be getting to the point in developed economies where more than 50 per cent of our access to banking services will be done from these types of smart devices. Basically, we need to have a more flexible method of delivering content to our customers.

We now have a myriad of systems and devices that provide different interfaces to our typical bank, such as teller and contact centre systems, ATMs, kiosks and in-branch systems, websites and m-portals, mobile apps and LCD panel displays. All these systems have access to the same back-end data, but generally the bank (or their agency) has to programme or design each interface completely in isolation from the others. This produces an inconsistent customer experience, and also means costs are duplicated because we are churning out multiple interfaces accessing the same data on different devices and platforms. Basically, the current channel silos are costing banks money needlessly.

So what's the solution? Well, there is a twofold element to attacking this issue for the bank. Firstly, the bank needs to invest in a mobile interface or a mobile applications team. The role of this team is to interface with the business and product teams to drive the functional and creative design of the applications to be deployed. Don't leave this to agencies; this is going to be a core skill of the bank in the coming years and it will be more cost effective to develop a conceptualisation team within the borders of the corporation.

Secondly, the bank needs to deploy more flexible interface and presentation layer systems. The presentation layer of information to the customer should be common across channels so that in the event that such content changes on one channel, it is immediately updated across all

Figure 6.15 Kyocera's flexible Organic-LED future concept phone, designed to fold up like a wallet (Credit: Kyocera)

channels where publication rights are active by the author. The bank needs to think like a publishing house or a news organisation in the way it handles content. Content needs to be delivered daily to customers on products and services, but the core transactional information and functional platforms need to be consistent in service and informational quality across every channel.

Mobile interface prototyping

Mobile applications development is a lot like any other application development, except you have to deploy information in a very limited space, so your effectiveness of informational distribution needs to improve.

Most banks would not have a clue as to where to start in respect of mobile application development, so they would simply approach some agencies, ask for pitch presentations and pick the one they like the best. But this is fraught with danger as most agencies are not banks and don't have the same customers as a retail bank. So they don't understand how the bank works, and they don't understand the demands of the customer. What they are very good at is taking a brief and executing creatively. So what should your brief for developing a mobile app to the agency contain?

First of all, it needs to contain a framework that the application can be developed upon. In industry terms, we call this an **Interaction Design (IxD) Template** or similar. This is not a hi-tech proposition. In fact, in most cases, IxD templates are decidedly lo-tech in their nature, being mostly delivered on whiteboards or sketched on pieces of paper. Thus, this is a process that can occur internally within the bank and result in a quality brief to the agency. This can then lead to an application that has both business potential and fills a customer need, rather than being an application that is just functionally workable or creatively exciting.

Prototyping is the method of iterative design, but it starts with a simple, yet very powerful tool or approach which is known as "sketching". Sketches are simple representations of the application interface that can be put in front of customers, product teams and business stakeholders for feedback and improvement. When we look at prototyping an application for a mobile device, we are looking at the following core elements:

- Structure and foundation of the application (core content or purpose).
- Behaviour (functionality and flow).
- Presentation.

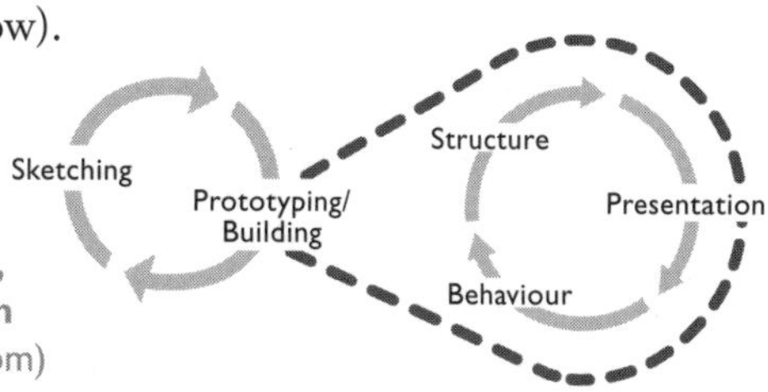

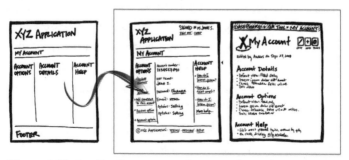

**Figure 6.16 Structure,
Behaviour, Presentation**
(Source: BoxesandArrows.com)

We start with the core objective of the application, the audience it has to satisfy, and the results it will achieve both for the customer and the business. We can brainstorm this with our customer advocacy team (probably call centre and branch personnel) and with the product teams or propositions teams responsible for customer interfacing.

Once we know what foundation the application has, we start to build on this through successive iterations, discovering the flow of the application and how we will get to the end goal. This process will probably start internally, but as we begin to develop successive iterations, we will put the sketches in front of real customers for feedback. Of course, staff are customers too, and we ought to use them. Once the sketching process is complete and we have a lo-fi prototype, we can move towards more formal prototypes. These finished prototypes can be annotated with technical details and then handed to the agency to build creative prototypes with a strong foundation. Why do this? Well, it saves you creating an application no one will use and gets the foundation right in a very inexpensive and rapid fashion.

Figure 6.17 Application prototyping — an iterative process
(Source: BoxesandArrows.com)

Presentation layer architecture—losing the silos

Various technologies are available to look after this multichannel publication approach, but most significantly they are:

- business intelligence and management solutions;
- messaging architecture;
- Web based protocols (XML/XHTML, CSS, J2EE, Ajax and RSS); and
- mobile protocols (such as SMS, MMS, LBM).

The key to utilisation of the mobile device or app phone of the future is not just a flexible "presentation" layer in your bank's IT architecture, but the ability to leverage intelligently this platform where the customer is concerned (Figure 6.18). As discussed, we are seeing a shift in consumer behaviour, but this means customers expect to be provided with the best financial decision options available as a result of their relationship with their financial services provider. In other words, their loyalty to the brand of the bank should be rewarded by impactful benefits, delivered in real time. The mobile device is an excellent tool in the bank's arsenal for satisfying this changing behavioural model.

Let's take a simple credit card transaction at a retail store. Let's say that this month in Hong Kong, HSBC is having a special offer for its customers who shop at the Lane Crawford store. Traditionally, the bank might send me a direct mail promotion for this "special offer". However, three weeks later when I actually walk into the retail outlet to look at a new three-piece suit, I've completely forgotten about that direct mail promotion. Given the ticket price on the suit, it's likely that I will use my credit card to pay for the purchase. As to which card I will use for the purchase, I'm typically going to use the card that has the lowest balance on it today, or some other sort of similar logic. Thus, if I have three or four cards, the chances are high that HSBC is going to miss out on my business despite the direct mail promotion.

The mobile device married with an intelligent POS (point-of-sale) system, however, could offer the bank a unique opportunity. If, when I walk into the Lane Crawford store, my mobile or the POS display reminds

Figure 6.18 Typical multichannel architecture configuration

me that there is a promotion at the store with a 15 per cent discount this month for HSBC's preferred customers, my likelihood of using that card will be much, much greater than it is today. Additionally, this sales message, which if caged in a promotional direct mail would likely be seen as intrusive or trying to "sell me something", suddenly becomes a service—my bank is offering me a great deal because I am a valued customer. The sales message becomes "service-selling" because it is delivered at the point-of-impact. We'll talk more about this in upcoming chapters.

"Early evidence suggests location-based advertising yields significantly higher conversion rates with direct response modes, such as click-to-locate and click-to-navigate, compared to non-location-based advertising."

Dominique Bonte, ABI Research

Revenue and Savings Opportunities

Like online banking, mobile banking is a "sticky" service that can increase customer loyalty and profitability. Although, like any channel, mobile banking introduces some operational costs, it is by far the lowest-cost bank channel on a per transaction basis, averaging about 8 cents per transaction versus $4 for the branch and $3.75 for the call centre.[28] Clearly, institutions can reduce costs by converting offline customers to the less expensive mobile channel. All the indications are that adoption of mobile is speeding up incrementally. Apart from those examples we've already covered, there are a myriad of global examples of mobile banking and commerce well underway:

- Japan has more than 40 million handsets in use that are capable of making payments at the point of sale.[29]
- Today, South Koreans can check their bank balances in real time, make bill payments, and pay for items in the check-out lane using their mobile devices.
- Absa Group Limited, one of South Africa's largest financial services firms, signed up more than a million mobile banking account holders, representing more than 25 per cent of their total customer base, and twice the number of customers accessing their banking from the Internet in a three-year period.[30]
- State Bank of India, Bank of India and Union Bank of India are going to introduce new mobile services in the nearest future. Mobile banking was launched in July 2008, with functions for account-to-account money transfer, utility bill payments and account-related queries. Some 17.5 million

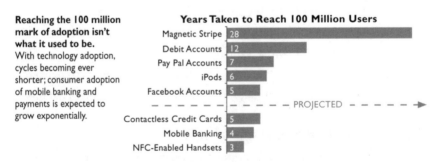

Reaching the 100 million mark of adoption isn't what it used to be.
With technology adoption, cycles becoming ever shorter; consumer adoption of mobile banking and payments is expected to grow exponentially.

Years Taken to Reach 100 Million Users

Magnetic Stripe	28
Debit Accounts	12
Pay Pal Accounts	7
iPods	6
Facebook Accounts	5
— — — — — —	PROJECTED — — — — — →
Contactless Credit Cards	5
Mobile Banking	4
NFC-Enabled Handsets	3

Figure 6.19 The risks and opportunities in a mobile commerce economy, 2008 (Source: TowerGroup Projections and First Data)

customers of Bank of India are expected to subscribe to the
new services over the next few years.[31]

• BofA has grown to have more than 3.5 million mobile
banking customers today, amassing 300 per cent growth
between 2007 and 2008 alone.

Table 6.3 is a checklist, if you like, of core revenue and cost savings
opportunities presented as a result of mobile channel initiatives and
developments. The cost savings are a given, if you work on channel
migration as an overall strategy for the bank. The rest require specific
initiatives within the bank and the application and multimodal access
approach to the mobile channel.

Table 6.3 Revenue opportunities for mobile banking services

Transactional Revenue	Acquisition Revenue	Subscription-led Revenue	Cost Savings
Mobile payments	New product acquisition	Monthly fee for daily feed	Reduction in branch, ATM and call centre load
Mobile remittance	Drive to Web, branch lead-gen	Royalty and card usage offers	Decrease in cost of acquisition
Virtual NFC debit/credit card	Up-sell, upgrades	Location-based usage offers	Decrease in net customer cost

A good example of integration and convergence is simply the ability to
work out which part of the day-to-day interaction with the customer can
migrate to the mobile. The earlier discussion on what goes into a typical
mobile application is a good start. But think about this. A customer goes
to an ATM and withdraws cash. We sync the new balance of his savings
account with his mobile phone as soon as the withdrawal is complete—
in real time. Thus he doesn't have to worry about putting in his card and
checking the balance again. We've saved a transaction and he now has an
up-to-date account balance on his phone—the number one requested
service typically through a call centre or ATM. Therefore, the cost savings
are obvious. Plus the service experience is better.

Steve Townend
CEO of MoBank discusses the future
of mobile banking as a service, not an
application

AUTHOR'S NOTE: MoBank in the UK is one of a group of third-party innovators that is taking on the banks at their own game. The banks may feel threatened by the emergence of players such as MoBank, but the fact is that if banks were more active in the same mobile space, then such a significant threat would not exist. Banks are just moving too slowly in innovating around the channel. As such, MoBank and other third-party developers have capitalised on this gap in customer expectations and needs. Below is an interview with Steve Townend, CEO of MoBank.

What is MoBank's long-term objective?

The long-term objective is to create a banking alternative. The wording is important—it is not just another bank; it is something different. It is what a (retail) bank should be. I want to use the attraction of technology to enable people to control their finances in a way that is unobtrusive, simple, relevant and intuitive, with no hidden traps. Leading with the mobile channel is a manifestation of all those things. Over time, I expect the channel to become almost invisible, allowing our members (not just customers) to simply transact—not just wherever and whenever they want, but also however.

I even envisage a high street store or similar, based on the Apple iTunes store model—an interesting and exciting place to go.

We want to be known for the way we deploy, make relevant, design and integrate technologies to help our members feel more confident and free to get on with their life. Branding is important too. We are getting a reputation for being irreverent, but transparent; on our membership side, creative, relevant, safe and inclusive. I feel we have an opportunity to become the way retail banking is, to really shake up the status quo and would like to consign banks to be manufacturers, the power to be in the hands of distributors such as ourselves. This will really create a truly competitive landscape and give customers real choice.

Why MoBank, why not the banks?

Banks have cost cut to the extent that most of their infrastructure is shot and the demands of their shareholders for short-term profits renders them almost impotent when it comes to innovation—the banks get rewarded for sustaining the status quo. Innovation is not simply repackaging an existing product in a new way.

Just look at how these banks reacted to First Direct and Egg, notwithstanding the fact that neither has really fully exploited the opportunity. Internet banking today is still pretty insipid, and telephone banking and IVR are reduced to a service that most customers still avoid as a bad joke.

At MoBank, we represent a future that is unsullied and unaffected by the past. Consumers, in general, want to believe in a new and better way. They are fed up with empty promises and being exploited. The way we are set up means that we do not need to behave in the way banks have in the past. We're not constrained, we don't need our customers to get into debt to make money, there are no hidden charges. We don't have legacy infrastructure to maintain. And we believe in our values. Banks are too biased towards creating shareholder value, rather than creating value for customers.

How do you think the smart device or app phone is going to change the way we do banking and handle payments over the next three to five years?

Absolutely, there will be a merging between the PC and the smartphone. Worldwide, the mobile Web will be way bigger than the wired Internet. The mobile will become a digital cheque book, the way you monitor and make financial transactions. NFC (Near Field Communication) technology will ensure that customers will be able to make payments in the physical world via mobile as well as the virtual. I would not totally discount ATMs either, I think they will become more like kiosks with more facilities, but these will become recharging stations for mobile payment devices, etc.

I know we can get carried away with technology, and where I come from there is a saying "there's nowt so queer as folk!" i.e. customers may not want all this stuff all the time. I think there still will be a place for branches and for cash. I think physical credit cards though will become a thing of the past. The fact that so much of our life will sit on the phone, not just financially, but, for example, the way you start your car, the way you order your TV programmes, manage utilities, pay for content, etc. It won't be magical or amazing, it will be just like switching a light on— all pervasive.

You've noted in your blog recently the culture of Londoners in respect of their travel time and the use of the iPhone device, etc. How do you think these sorts of behavioural changes of consumers might affect purchasing behaviours in the near future?
I think the first purchases to migrate to the phone will be those that represent instant gratification and those [that] need to be done in a hurry or in real time. It is about the ultimate convenience. Initially, it will be assisted by hot deals, special offers and exclusives. Purchases that are of a relatively small value and do not require too much research will be the easiest to sell. For the same reason that complex products like cars have never sold over the Internet, simple products like pizza, gifts, flowers, movie tickets make for great potential on your mobile because you don't need to touch or feel them to make the purchase decision.

Personally, what I have found is not only do I purchase things on the move in what I term "interspace" (the time in between social, home and work life), but I also use it at home. For example, I was reading *The Times* literary supplement recently, and just ordered and paid for three books using my phone and it took all of less than a minute to do it. I think like me, customers will quickly latch on to one or maybe two typical facilities, for example, I always buy books this way or get my theatre tickets through my mobile. This is what we are already seeing in respect of transaction behaviour on Mobank. Exciting times!

What, if any, are the barriers currently to better banking via mobile?

Security is no doubt the biggest fear both for consumers and bankers alike. The fact that in reality today it is now more secure to do banking over a mobile phone than through a PC makes no difference to perception. Accessibility is an issue too. However, as MoBank becomes the ubiquitous service platform, this gets over the hurdle of accessibility.

For MoBank, it doesn't matter who you bank with, what mobile operator you are with, what phone you're on or what payment vehicle you want to use. Just simply that we allow you to get to your money and leverage it whenever you need to, wherever you need to. The added advantage of our embedded loyalty programme where you are rewarded for going mobile with special deals and offers should also reduce the negative impact of perceptive barriers. Customers get over these issues pretty quickly. Look at Amazon, airline tickets, and so on.

That's great. Thanks for the insights, Steve.

KEY LESSONS

If you had asked most bankers five years ago when they thought that mobile banking would become mainstream, they would likely have told you "not in my lifetime". Yet, in the last five years, that is exactly what has happened and now banks everywhere are talking about mobile banking.

The problem is that unlike Web pages and other interfaces, mobile phones have small screens and are pretty limited in what they can do. So how can this device realistically become an effective channel for transactions and payments?

Bankers need to start treating mobile as a serious competitive advantage. New competitors in markets such as the UK, Japan and Brazil are exclusively mobile banks, and they are winning customers. Accessing my bank while on the move has just become a very basic requirement for bank customers.

What will the future hold? Well, mind-blowing emerging technologies, a raft of new application capabilities, and increased mobility of customers make finding a solution to the mobile banking conundrum a core skill for the CEO of a bank today!

Keywords   Mobile Payments, iPhone, Wireless, Mobile Banking, Unbanked, Remittances, Contactless, Marketing, Promotion

Endnotes

1   www.worldgsm.com data, 2004

2   Average Teen Texts, www.nytimes.com/2009/05/26/health/26teen.html

3   iPhone™ is a trademark of Apple Inc, and Android™ is a trademark of Google Inc.

4   The Industry Standard, 2009

5   Motorola Media Centre

6   "The 50 Greatest Gadgets of the Past 50 Years", Dan Tynan, *PC World*, 24 December 2005

7   Nokia 888 concept designed by Tamer Nakisci (Nokia design award). The other two concepts are shown here with the permission of their respective trademark/copyright owners. Caps concept by designer Jean-Jacques Chanut.

8   *CNET News*, 19 November 2008. Q&A: Kurzweil on tech as a double-edged sword, Natasha Lomas

9   iTunes™ is a trademark of Adobe Systems Inc.

10 Losing iReligion, gedblog, http://gedblog.com/2009/09/28/losing-ireligion/

11 UserCentric.com. "Early Adopter iPhone User Study Identifies Baseline Issues with iPhone Interface", 12 July 2007

12 148Apps.biz. App Store Metrics, Apple Press Release (Apple.com)

13 The HTML5 is already being used in applications such as Google Wave and elsewhere, but won't start to be widely accepted until 2012.

14 See Wikipedia.org articles on WiMax, 4G, UMTS, and Spectra Efficiency of long-range networks utilising 802.11, 802.16 and 802.20 standards.

15 Time.com

16 February 2008: ICM Research found 16 per cent of Brits had changed banks (sample size 1,006) because they were unhappy. This is compared with the UK divorce rates of 45 per cent (Source: UK Office of National Statistics). See www.icmresearch.co.uk

17 UserStrategy 2004–09 (Asia, Middle-East, UK analytics/surveys of major retail banks), Alexa.com

18 Juniper Research forecasts total mobile payments to grow nearly tenfold by 2013.

19 Juniper Research: Mobile payment transaction values $300b globally within five years

20 "In South Korea, All of Life is Mobile", *NY Times*, May 2009, http://www/nytimes.com/2009/05/25/technology/25iht-mobile.html?pagewanted=all

21 World Bank

22 GSMA. gsmworld.com

23 Edgar Dunn and Company. edgardunn.com

24 *The Economist*, 26 September 2009

25 "Big Banks in Plot to Kill M-Pesa", *Nairobi Star*, 23 December 2008

26 "Philippines mobile phone-based microfinance bank set for launch", Finextra, 13 October 2009

27   Within the next 10 years, we will see e-paper and flexible OLED (Organic LED) devices becoming quite common, removing size restrictions.

28   McKinsey Quarterly and GSM Association

29   Julia S Cheney, "An Examination of Mobile Banking and Mobile Payments: Building Adoption as Experience Goods", Payment Cards Center, Federal Reserve Bank of Philadelphia Discussion Paper, June 2008

30   Brandon McGee, "Absa Reaches 1 Million Mobile Banking Clients", Mobile Banking blog, 23 February 2009

31   Bank of India press release, "Mobile and banking services integration", *ECommerce Journal*, 30 June 2008

7 ATM and Self-Service Banking —Convergence and Control

by contributing author, Michael Armstrong, former Senior Manager, Customer Propositions, HSBC Asia Pacific

Self-Service Banking—Where It All Started

THE first mechanical cash dispenser was developed and built by Luther George Simjian, and installed in 1939 in New York city by the City Bank of New York. Six months later, it was removed due to the lack of customer acceptance. The first self-service device of any note that was a commercial success was the Automated Teller Machine launched by Barclays Bank in 1967. That device relied on a prepaid token to retrieve envelopes with a fixed amount of cash within. From this relatively primitive beginning, ATMs have gone on to revolutionise the banking habits of most retail customers.

The ATM solved one of the biggest problems for a retail/commercial bank, that is, the distribution of cash for customer withdrawals since its mass launch in the 1970s in the US and most of the rest of the world in the 80s. The ATM celebrated its 40th birthday in 2007, and today 75 per cent of all cash in the UK is dispensed to consumers via the ATM. Cash machines are an essential part of most consumers' daily lifestyle.

The invention of the ATM meant that one of the biggest fixed costs in any retail banking operation—the branch—could be reduced through branch rationalisation (closures). In addition, the variable cost component—staff—could also be reduced. So, ATMs provided one of the biggest one-off saving hits for branch banking. The automation of one of the basic functions of a bank not only reduced costs, but also increased customer convenience, allowing access to cash 24/7. But this is where the story seems to have ended for ATMs.

While there was the initial big roll-out and the promise of further automation of services through self-service, the promise has never really been delivered. The early ATM machines allowed cash deposit, but in a rather unsophisticated way. Customers were generally reluctant to place cash into an envelope, insert it into a machine, and then wait for verification from the bank after the staff had counted it, maybe 12 hours later. Sure, ATMs have evolved in terms of their look, efficiency and speed, but their function, on the whole, is still concentrated on cash delivery.

What about the other self-service devices apart from ATMs? These include cash deposit machines, cheque deposit machines, stand-alone kiosks selling products, passbook update machines and instant balance machines. These can be either incorporated as part of a branch or as stand-alones. They allow instant recognition of both cash and cheques and online account crediting, as well as instant sale of products. We will take a look at the success factors for the various self-service devices.

So, while the initial concentration on self-service was cost savings and processing efficiencies, the drift then moved to revenue opportunities. As the efficiencies that were created by the introduction of ATMs plateaued, management needed to look at revenue opportunities. Banks started with the most obvious transactional features, kicking off with account transfers, bill payment and statement requests, but now are looking at the sale of products through the channel. There have been much heralded instances of banks selling products such as insurance through the ATM, but is that truly viable?

Self-Service—the Promise and the Reality

The drive for efficiency

The initial drive for banks to launch ATMs was to promote an innovative image in what was still a branch dominated world in the 1960s and 70s. There was little effort at migration of customers from branches to ATMs in an era of tight banking regulation in most of the world.

Regulation meant that banks were limited in what they could offer in terms of competitive interest rates, and with credit rationing, especially in housing loans, there were set quotas on what could be lent and at what level of interest rate. In such an environment, there was little incentive to be competitive. In effect, the banks were cross-subsidising their services, including branches, with the income earned on their lending and deposit services. With typically no or low interest rates offered, in fact, Australian banks would charge customers for the privilege of holding a current account, and with lending provided at high, usually government sanctioned, rates, the banks made very healthy margins on their lending business. These profits would then go to offset the costs of providing most other services offered by the banks, the biggest being the cost of running branch networks.

So while self-service devices were around in the 1970s, there was little incentive to encourage active migration to them as branches were still the main platform of service and were being well funded by cross-subsidisation from heavily regulated lending services.

Deregulation of the financial industry in the UK, US and Australia in the late 1970s and early 80s meant that financial institutions could no longer hide inefficiencies. With competition allowed in the deposit and lending markets, margins were starting to be squeezed and services had to be justifiable on their own account. This was especially the case in countries with large branch networks such as Australia and Europe. The largest fixed cost was branches and the highest variable cost was staff. When looking at the activities of the branch, the accountants found that a huge majority of the work performed related to the teller function which consisted of cash deposits and cash withdrawals. If you take out these two elements, you take out up to 70 per cent of the branch staff costs.

Interestingly in my earlier banking years in the late 1990s, I conducted a time and motion study on our branch network. I selected six branches in which to time activities, and staff were asked to record how many of those activities they undertook. With a set time for each activity, I could then work out the percentage of time spent on each activity. The purpose of the study was not to work out any grand strategic plan on which services should be migrated to other platforms, but rather a cost allocation exercise to ensure each product took its "hit" from the cost of running the branches.

The study showed that 10 per cent of branch time was spent on conducting credit card transactions, either selling them or accepting payments, or making cash advances through the teller council. The reaction from the head of the credit card department was to threaten to withdraw his product from the branch network—obviously it was not a really viable solution. What the results showed was that well over 50 per cent of all branch activities were still related to cash and cheques that could potentially be migrated to a cheaper platform, but were presently still being performed at the branch. Each one of these transactions that made up the 50 per cent load on the branch was essentially a cost. Recent studies in the Middle-East have shown that about 60 per cent of OTC (over-the-counter) transactions there are in fact either cash transactions or cheque deposits.

As an aside, and something we will discuss later in this chapter, the reason we still had such a significant minority of customers using the branch was that they still preferred to use a human for the transactions because of the **trust** they placed in receiving an **instant confirmation** of the transaction through a receipt with a teller stamp on it. So while banks had instigated massive expenditure on self-service, many customers still wanted the comfort of a teller stamp on their receipt especially for what they considered "critical" transactions.

Let's return to the ATM journey. With such large costs sitting there, the banks then embarked on one of the biggest customer education exercises ever attempted. How can we push people out of the branch and to use the ATM platform for their cash transactions?

Like most technologies, the early adopters readily accepted migration as they had confidence in the technology and saw the obvious convenience that went with transacting at a time suited to them and avoiding queues inside the branch. The carrot worked, but then ATM usage began to plateau in the late 80s and early 90s.

As of 2009 in Australia, ATM withdrawals accounted for around 63 per cent of the value of cash withdrawals, while the value of OTC cash withdrawals still accounted for around 26 per cent of cash withdrawals.

Forrester states that 69 per cent of bank customers in Europe use ATMs on a regular basis.

This number appears relatively decent, but considering ATMs have been around for more than 30 years, wouldn't you expect the number to be higher? To "encourage" more usage of the ATM channel, banks in Australia started to charge for OTC transactions, with some banks charging $4.00 for each branch withdrawal, while some chose to offer a limited number of "free" withdrawals before they started to charge.

In addition, banks began education programmes, especially with the elderly, to show them how to use the ATM. So while ATM has become by far the preferred channel for cash, and despite the "stick" of charges for OTC transactions, there are still a considerable number of people who prefer to use the branch for cash and are willing to wear a cost to do it.

It is a similar story in Asia. Surveys of both HSBC and Standard Chartered's customers based in Hong Kong, Singapore, Malaysia and elsewhere show that in respect of frequency of use, more than two thirds of customers will choose the ATM for cash withdrawal on most occasions rather than the branch.

Here are other factors of interest with respect to ATM usage or choice:

- The most preferred location for an ATM was near a metro station or in a shopping mall, not at a branch.
- Eighteen per cent of respondents said they might ring their bank or check the bank's website for the nearest location. Google Maps anyone?
- Ten minutes' walking time is the maximum most customers will expend to get to an ATM.
- The three most important factors in choice of ATM location are:
  - safety and security,
  - proximity to public transport or shopping mall; and
  - proximity to home or place of work.
- ATM is used at least five times more frequently than a branch.
- Customers also indicate that they would use EPS, EFTPOS or POS implementations with cash out more frequently if available.

What About the Revenue?

Any decent financial controller working in a bank will see that the use of ATM to distribute cash achieved considerable efficiencies, resulting in substantial cost savings for banks. With the vast majority of customers now using the channel for cash, what else can a bank do to maximise its investment in ATM machines?

The first step was to incorporate additional **transactional** type services, including account-to-account self-transfers, statement requests (these would then be posted to the customer at his normal address), third-party transfers and bill payment.

The second step was to search for **revenue opportunities**: "What can we sell to the customer while we've got them standing here?" It is easy to understand the reasoning behind this. In Hong Kong, a few banks started to offer subscription to IPOs through their ATMs, while in other parts of the world ATMs started to offer concert tickets, accident insurance policies and other commodity type offerings. While it was technically possible to offer these products through the platform, is it really what the customer wanted? Is it how they envisage an ATM to be used? To most people, it was still primarily for cash, but could they be sold something else through the platform? In our next section on pages 243–249, we'll look at a study on customer reactions to self-service devices.

Interchange and surcharge fees

The primary revenue potential from ATMs came from the ability to charge a transaction fee for using the machine. Banks traditionally charged one another (interbank transfers) when a customer used a "foreign" ATM. This fee was charged to the customer indirectly through their own bank. The customer's bank then had the choice of either absorbing the fee or passing it on to their customer.

Let's use the Australian ATM system as an example of what has evolved. When ATMs were launched, each major Australian bank went out on its own to establish a competitive advantage based on the number of ATMs it could offer its customers. The banks felt they could individually provide adequate customer service through their own proprietary network.

At the same time, the building societies (pseudo-banks that took deposits and made loans) decided to network their ATMs into a network called "CashCard".

The reasoning behind the CashCard network was that while there were many building societies, each on its own was insufficient to offer its customers sufficient coverage nationally. As a combined network, however, they offered greater coverage than even the individual banks. The banks responded with a series of bilateral agreements whereby the Big Four banks paired up to share their respective networks. Customers using the "other" bank in the network were not charged for using a "foreign" ATM as the banks wanted to promote their ATM coverage.

The interconnection between the different banks was made possible by charging a fee every time a "foreign" transaction was made, i.e. a customer from another bank using my bank's ATM. Historically, the actual charge for the transaction was hidden from the customer who had to wear a set fee that appeared on their monthly statement.

In an enquiry into ATM practices, the Reserve Bank of Australia (RBA) found that the average fee charged by the Big Four banks was $1.00, whereas the actual cost was closer to $0.50. The reason was that there was little competitive pressure to change the fees.

What drove this fee structure? Firstly, the banks had little incentive to renegotiate rates as they had little choice on which other bank to pair up with. If they walked away from an existing arrangement, other viable partners were very scarce, so it was easier to stay in the cosy existing relationships. While this suited the banks, their customers suffered because they had to absorb both the fee and the bank's "operating" margin. Secondly, there was no real pressure from customers who were charged the same fee by their bank, regardless of which ATM they used. This meant that ATM providers had no incentive to compete on price.

The solution that the RBA decided on was to allow "direct charging". This system scrapped the interconnection fee charged by their own bank and allowed each ATM provider to charge a fee at point of transaction. This way the pricing is explicit, and competitive forces should mean that ATM providers price according to demand and supply. This should also create an incentive for third-party providers to enter the market as they no longer need to be part of a bilateral arrangement and can service any cardholder for a set fee.

Since March 2009, ATMs in Australia have been "direct-charge". This has incentivised one of the big players, National Australia Bank (NAB), to tie up with a third party ATM provider, rediATM, and in the process they have effectively doubled their ATM network availability. In this case, NAB customers are not charged for using the redi network as NAB absorbs the cost. It has been estimated that over $500 million in direct charge ATM fees are incurred by consumers in Australia each year. ATM direct charge fees in Australia range between $1.50 and $2.20 for cash withdrawals, and $0.50 and $2.00 for balance enquiries.[1]

In the US, the shift to imposing direct charging or "surcharges" occurred in 1996. The effect was to turn the ATM into a profit centre earning income from charging surcharge fees to card customers. The number of ATMs surged as third-party providers sought to build third-party networks with the ability to earn fees. Interestingly, it seems that the surge in ATMs in the US resulted in a reduction in ATM income per machine as customers judiciously avoided these fees by using their own banks' machines.

At the beginning of the ATM era, banks provided ATMs as a point of differentiation. They then hooked up into networks which made the ATM ubiquitous and a commodity. With the advent of direct charging, or surcharging, third-party providers entered the industry charging variable rates to use their ATMs. With the income provided by ATMs reducing, perhaps banks will again start to see these as a customer service tool rather than just a profit centre. There are other revenue opportunities that present themselves for the more adventurous of banks, however.

The "Other" Self-Service Devices

The current generation of instant cash and deposit machines provide instant checking and verification, and provide a customer with a receipt. In the case of cash deposits, instant balance updating is even possible. In Hong Kong, the advent of these instant cash and cheque machines has seen transaction volumes drop by more than 30 per cent in branch networks in the last few years. To hark back to Chapter 3, the positive business case in usage of these devices is that it enables branches to focus less on net-cost transactions (cash deposit/withdrawal, cheque deposit/cashing), and more on sales and service capability.

The more complex question is how to make revenue out of self-service devices today. Is it as easy as installing an ATM, "build it and they will come"? The problem with selling a product is that there has to be a need that requires a solution. For the ATM the need is very obviously cash, and cash is easy to sell!

Cash is used as a medium to buy anything. Its need is universal and is so obvious that marketing does not even have to sell it. So long as the ATM is visible, it will have customers. Now, as we said earlier, the bank's financial controller wants to start *making* real money out of self-service because it means the bank has fewer people in branches and a wider reach without having to invest in full service branches.

Now let me tell you another story...

In Hong Kong, we had established a highly successful Internet-based offering built on commodity insurance products. Here are the reasons for its success:

- Products were easy to understand and tallied with readily identifiable needs.
- Only essential questions, and as few as possible, were asked.
- There was instant payment solution via credit card.
- There was instant approval and the customer could print out confirmation of coverage immediately.

We had great success with travel insurance because it fitted the criteria listed above. The need was apparent—if you go on holiday you need insurance. All that was required was essentially the dates of travel, the applicant's name and the credit card number. About one year after the launch, the website was handling over 50 per cent of all applications for that policy type, and so we thought we had done a great job.

Then one morning, I saw an article in the paper from a rival insurance company proclaiming its success at selling travel insurance through self-service kiosks. These kiosks were actually "smartphones" that were really self-service kiosks. It had a large touch-screen LCD display that allowed many functions to be conducted through it, from buying tickets to finding the nearest convenience store, as well as making the humble phone call. The kiosk used Web-based architecture and had a built-in printer, which meant that receipts and tickets could be issued.

The company that owned the kiosks placed them in strategically high-traffic areas—train and bus stations—and places where people congregate, most importantly at the Hong Kong/China border. It was apparent that they did not see themselves as a phone operator, but rather as a multifunction kiosk. They made their money from renting out applications on their kiosk/phone. The model is similar to what Apple™ now does with the iPhone. They build the hardware and then allow third parties to sell applications that can utilise the functionality they have built.

Now, the insurance company approached the kiosk vendor and said they wanted to sell simplified travel insurance for people going to China. They were only interested in the kiosks located in the train station where trains left for the border. (On busy weekends, the number of people who travel that route can be in the hundreds of thousands.) They also requested that their "app" be on the kiosks.

Next, they implemented a very simple and focused application process. They only offered one type of travel insurance for people travelling to China. The customer was only required to enter his Hong Kong identity card number and indicate the number of days of travel. There was no need to provide his name, contact details or address. Payment was made using the Octopus card, a contactless smart-card carried by 98 per cent of the Hong Kong population for their transport needs. The machine would print a receipt confirming the purchase. The entire transaction could be completed in about 15 seconds.

These people had really ticked all the boxes on how to sell in a self-service environment—the product was simple and focused, highly visible, easy to apply for, easy to pay for, and you get instant confirmation.

Well, if they could do it, why couldn't we? So HSBC approached the same vendor. We knew that the application form we were using on the Web had to be simplified, but we could not get away with entering just an ID number due to compliance restrictions within the bank. Nor could we do a contactless one-second payment swipe using the Octopus card; we still needed the customer to enter the 16 digit credit card number.

Nonetheless, we decided to hire 10 kiosks at Hong Kong International Airport—lots of traffic and surely there would be people who had forgotten to get insurance! But with an airport so large, it was very difficult to get all the major points covered. In the end, we placed some kiosks in the passenger check-in area, as well as in the area after passengers clear immigration. We created colourful banners advertising the travel insurance product and displayed them around the kiosks. We thought we had the elements for success covered—a simple product, a simple application process and a great location.

The results were disappointing. Our sales measured in the single digit for most days. As a percentage of overall sales, the channel was statistically insignificant. What happened? Through surveys with our target customers, we found we had violated the rules we set out earlier. While our product was good, there were deficiencies in the following areas:

- **Location.** Too disperse to gain a critical mass of customers.
- **Application.** Still too complicated. There were too many buttons to press which increased the probability of errors. The customer would give up out of frustration and also to avoid raising the ire of people standing behind him waiting to use the kiosk.
- **Payment via credit card.** Took too long compared with a stored value smart-card.

Our kiosk venture failed. The application we had was perfect for a Web-based application; we knew that because of the tremendous success of that channel. There is a big difference, however, between sitting in the office or at home completing an application on your PC, and standing in a busy airport trying to punch your credit card number while keeping an eye on your bags and another on an unfamiliar interface.

There was another reason for the failure of this project—travel insurance could also easily be bought via mobile phone. Just call the insurance hotline and get instant coverage. In terms of total application time, the phone would have been by far the quickest way to get covered when out at the airport.

But where customers had the choice, they still preferred buying travel insurance over the Web. There is a certain degree of control in being able to compare premiums and the insurance terms with that of other providers. Moreover, one need not wait for confirmation through the post.

The key lesson learned from this experience is never to take the customer for granted, and never assume that what has worked elsewhere for someone else will work for you.

What the Customer Really Wants Out of the ATM

As has been stressed in the previous chapters, the key change in banking is to move away from a process oriented culture to a customer-centric focus. While banks often talk about the importance of the customer, whether they have really made the shift to such an imperative is debatable. We have spent a whole chapter looking at how the Internet is underleveraged and the customer seems to have been largely forgotten in the singular pursuit of cost reduction, so looking at the customer side of self-service produces a similar picture.

The key problem with the invention of the ATM, similar to that of the Internet, is that it was invented by visionaries, but then built by IT and run by purchasing and premises departments. No one could really decide if the ATM was a computer, a part of the branch, or a piece of furniture. A useful question to ask your bank is: **"Who decides where an ATM is located and what functionality does it provide?"**

In some banks, such decision making is decentralised with local area managers determining where ATMs should go. In others, a centralised function will determine where to put ATMs based on the size and location of the branch and its customer population—rather like how McDonalds decides where to put its stores. What banks really need to do is use analytics to determine where the need is, and the type of machine required and its features or functionality. Do all machines need to be the same and offer the same functionality?

So, what does the customer think and want as they approach an ATM?

At this point, I'd like to cover some research conducted in Hong Kong, Singapore and the UK over the period 2003–07. The research looks at customers' reactions to self-service devices and their functionality, and also at their likes and dislikes. It is practical and provides insights for maximising the potential of the channel.

The survey consisted of questionnaires, street interviews and usability simulations in an ATM "lab". We don't pretend that this is the entire, definitive story, but the snapshot of opinions garnered from the research are extremely valuable insights useful for all bankers. In keeping with this book, let's show the reality—what the customer thinks.

In a single word—CASH. The preceding discussion about the other potential uses for ATMs to sell products just doesn't wash with a large majority of participants. Historically, the most popular transactions are cash withdrawals, account balance, cash/cheque deposits and funds transfer. The research found that customers are generally in a hurry when they go to the ATM. Not only do they want to get on with their business, but they have the added pressure of people behind them waiting in the queue. But on almost every occasion, the customer's primary goal is first and foremost to get cash—that is their focus as they step up to the machine.

As a test of how focused the customers were when using an ATM, the respondents were asked immediately after the transaction to recall what was being displayed on the ATM screen at the time of the transaction. (On most Windows-based ATMs, there is the provision to promote a product or service on the screen while waiting for a customer or during processing).

The result—94 per cent of the customers could not remember seeing any advertising related messages at all. So extremely focused were they that anything not specifically related to the task was filtered out. This indicates that the recall rate of current forms of onscreen advertising is very low. The challenge here lies in converting what is essentially a processing platform—the provision of cash—into a sales or marketing platform.

Seth Godin, in his book *Permission Marketing*, explains that such "interruption-based" marketing or broadcast advertising is increasingly being filtered out by consumers. Have you ever watched a movie on TV and find you have to turn the volume down when a commercial comes on, and then pump up the volume again when the movie resumes? Well, this is because advertisers (and networks) crank up the volume on their commercials. They have figured out that when the ad breaks come on, people often get up to do something else. If they pushed the volume up louder, people might still hear the TVC, and therefore they might still remember the brand when they are next at the local supermarket.

One of the reasons TiVo has been such a huge hit in the US is their ability to filter out these annoying advertisements. It is also a further reason downloads of TV series has become so popular (both legally and illegally). Banks should remember this, because an ad on an ATM should not be intrusive and stop the customer from completing the task. If advertisements or promotions are complementary and integrated smartly into the experience, they may still work—but this requires some thinking and interaction design.

I have a need for speed

Customers were also questioned about their interest in having personalised messages on the ATM machines, such as status updates or loan approvals. The response was lukewarm; it appeared that customers may be interested, but not at the cost of longer waiting or interaction time. They stated that if the service was introduced, it would have to be presented in an unobtrusive way that didn't add to the overall transaction time. While most customers were hesitant about having personalised messages, the one thing they consistently demanded was an increase in speed of transactions.

Through the usability lab testing, customers identified areas for improvement in both the transaction processing time and in the interface efficiency. Though the processing time is subject to technical limitations, it was interesting that customers offered suggestions on how the screen flow could be improved to shorten the time to complete the transactions.

Let's look at the cash withdrawal flow as an example. Banks generally list the accounts the customer has with them using account numbers that can be up to nine digits long. Customers, however, would prefer to use terms such as "current account", "line of credit" or "savings account". Each time they reached a step with account numbers on it, it slowed them down perhaps only three to four seconds, but this was perceived to be a long time. In fact, as a result of this research, HSBC engaged the author's company, UserStrategy, to assist with optimising screen flow on its ATM machines. With some simple usability tests, we found many opportunities to reduce the number of screens and steps involved.

Moving non-cash transactions to a dedicated (non-ATM) machine

As we noted, the top transactions are cash withdrawal, balance enquiry, funds transfer and cash deposits. Further down in usage are bill payment and statement request. In the interests of efficiency, since the top two transactions of cash withdrawal and balance enquiry make up over 90 per cent of all transactions conducted, why not put them on their own machine to speed up transaction time for customers who just want cash?

Would customers be willing to use a basic machine that only gave cash? Customers were queried about their attitude towards the introduction of a "quick cash" ATM, a machine that would only have withdrawal functions. The response was yes, as long as there were other machines at the same location that could provide the other services required.

So, while 75 per cent of respondents liked the idea of a quick cash machine if it could speed up withdrawal transactions and decrease waiting time, they also wanted access to other machines that handled non-cash transactions at the same location. Transactions that are considered candidates for such a shift might include bill payments, cash deposits, statement requests and cheque book requests.

Those in favour of having a quick cash machine or not opposed to having one accounted for 79 per cent and 85 per cent respectively. This is strong support and indicates that such a move would have a positive impact on customer channel utilisation. One reason for supporting a cash only machine may be reluctance to take too long a time. Indeed, focus groups showed that the awareness of other customers waiting in line and their impatience was a deterrent in conducting the lengthier non-cash transactions. Customers are also concerned about the possibility of shoulder surfing. As one customer explained:

> "I avoid paying bills using the ATM because it takes too long and other people waiting in line lose patience and stare at me."
>
> HSBC ATM Customer focus group feedback

When customers who want to pay a bill, make a deposit, or to a lesser extent make transfers, are faced with long queue times, they often abandon the task and search for a quieter location, or postpone their transaction, or go to a branch. Those familiar with online banking may choose to do the transaction later online if it is not cash-based. It appears that customers would be more comfortable doing non-cash transactions on dedicated machines. This is related to the perception that taking more time at the machine, increasing other customers' waiting time, would be justified because the machine would be dedicated for that purpose and not for the vast majority just wanting to withdraw cash. The recent addition of barcode readers on ATMs for bill payment is an extremely positive step and sure to pay dividends.

Some insights from UserResearch
Withdrawing cash

In the usability tests, all the participants completed the withdrawal task with ease. Customers are very familiar with withdrawals, as it is the single-most frequent transactions they conduct in real life. Most ATM users have gained an "expert level" of competency conducting withdrawals and do so very efficiently, barely reading on-screen menus or instructions.

Some users did have problems, however, when it came to choosing the account to withdraw from. Not every customer is familiar with the numeric format differentiating savings accounts from current accounts. One user expressed confusion at the screen on which he is prompted to select the account. He did not know which one was his savings account, and which one was the current account simply by looking at the numbers. As a result of this feedback, ATMs for HSBC in Hong Kong now show the account name and number (the first and last digits only for security purposes).

For many ATM menu flows, a withdrawal transaction is completed by asking the customer to remove their card from the machine. Unlike many other transactions, the user is not asked whether they would like to continue and do another transaction; the screen prompts the user to remove the card and take the cash. Although users indicated that they would like to be able to choose whether to carry on, this might not be feasible for security reasons as the feature prevents cards from being left behind in the machine after a withdrawal. User feedback prompted a further change to the completion of the withdrawal task flow. (See the case study on the interface redesign on pages 250–251.)

Transferring funds

Transfers were the second most common transaction after withdrawal (excluding account balance enquiries). During testing, this task was conducted without any major hurdle. The only comment users had about making transfers is that they would like to be able to confirm the payee's identity was correct when they entered the "TO" account. While this may not be possible for security reasons, a final confirmation screen could be provided before the customer executes the transaction.

Paying bills

Users found bill payment generally straightforward. However, practically all users made a mistake when looking for the correct category to which "water" company bill payments belonged. Almost all users entered the "utilities" section of the bill payment menu. The problem in the case of the Hong Kong ATM structure was that water utility was a government run facility, so HSBC had categorised this under government payments, not utilities. This was the wrong classification for customers, and it illustrates the usefulness of getting customers involved in the definition of menu nomenclature.

Depositing cash

Customers often expressed the view that deposits are difficult to conduct on ATM machines. Observation field studies showed us that users encounter many difficulties throughout the procedure.

Users new to depositing cash had difficulty following the instructions on screen, and rarely followed them properly. The instructions prompted users to take the transaction receipt, place it in a deposit envelope (from a draw), seal the envelope, insert the envelope into the deposit slot, and press "enter" when ready to deposit the envelope in the slot. Many users did not take the receipt, some did not press "enter", and some were stuck looking for the slot.

One user in the test scenario timed out and had his deposit attempt cancelled. On his second attempt, he inserted the envelope with the deposit receipt from the previous transaction. The research here showed once again that if you provide too many steps, no matter what on-screen instructions you give, customers will have difficulties.

"The procedure for the deposit is rather complicated, [it] should be simplified"

A *customer*

The lack of good design also plays a large part in creating difficulties for a task such as cash deposits. The deposit slot remains a challenge. Some users could not find the slot for the deposit. Others had difficulty understanding which way to insert the envelope in the slot. One user struggled to place the envelope in the slot at an angle but the envelope would not go in. Another was gutsy enough to put his fingers inside the deposit slot to make sure it was open. A different slot design with a slight funnel effect would make the deposit direction more obvious, and would help guide the envelope to the correct angle. As for the envelope, one user discovered it had a self-adhesive tab only after she licked the envelope and found that it sealed partially.

Luckily, newer ATMs accept and count the bills as you insert them thereby removing the need for envelopes and receipts—altogether a much better solution.

The key lessons from UserResearch on self-service transactions are these: make it easy and intuitive, make it fast, and provide the transactions most likely to be used. A simple rule to remember is this: **If you have to provide a set of instructions on screen for a customer to know how to complete the transaction, then it is simply TOO COMPLICATED!**

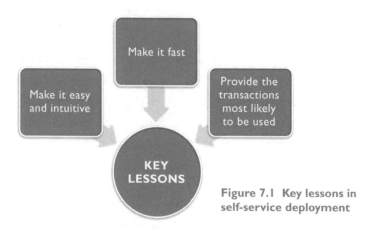

Figure 7.1 Key lessons in self-service deployment

Moving beyond just usability improvements

What you see in the case study is just the first stage of the general improvement exercise in the usability of HSBC's ATM network interface.

In December 2009, HSBC launched a new ATM interface schema, this time with an even more significant improvement in usability and efficiency through touchscreen technology. The ATMs incorporate personalisation functions and reduce screen flow dramatically—cash withdrawals are down from a total time of more than 30 seconds to just under 20 seconds. For bill payment, screen density is reduced from eight pages to just three screens by using a bar-code reader for the bill/invoice. Recurring transactions and bill payments can be saved for later recall to improve speed further.

The devices also enable more user-friendly accessibility for elderly customers. According to Hong Kong newspaper *The Standard*, the new ATMs are to be installed at more than 700 sites over the next three years.

CASE STUDY

The HSBC Experience

ATM Screen Flow Redesign leads to 23 per cent improvement in speed of withdrawal transactions

A qualitative usability test was carried out on the OS/2 based and WinXP based ATM platforms to discover any user experience issues. The results determined that certain key transactions could be simplified to enable faster transaction resolution times, and a better user experience. Below is the comparison between the original screens and the new screen flow with comments about the change

OS/2 VERSION   NEW WINXP VERSION

COMMENTS:
Notes were made more prominent, as was the appropriate customer service hotline.

Prompts language was simplified and the cash withdrawal advice option was moved to the final step.

The Account selection was moved to second step.

The "Amount to Withdraw" and "Other" or "Key Amount" options were combined into a single entry screen.

OS/2 Version

New WinXP Version

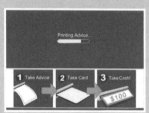

COMMENTS:
Account name was added to the selection options because many customers didn't recognise numbers alone.

Simplified language and options available at the close of the transaction.

"Withdrawal Accepted" and "Print Advice" screens combined into one.

Screen flow here shows much better feedback for customers as to what stage of the process they are at. If the user selects balance option, the screen displays the balance on-screen.

The Future of Self-Service

Other than distributing cash, the ATM and other self-service kiosks lend themselves to two potential possibilities in the near future:

1. Distribution point for other services that are non-financial.
2. Hub for financial transactions apart from straight withdrawal from a bank account.

We have previously discussed some of the potential product sale opportunities such as simple insurance products, but what of other non-financial uses?

ATMs as distribution point for non-financial services

One reason banks have the opportunity to exploit the uniqueness of the ATM proposition is the sheer number of ATMs/self-service devices out there. There must be some other use that we can put to the machine with its Windows screen display, keyboard, dispensing capability and network connectivity. The obvious item that springs to mind would be products such as concert or movie tickets. Perhaps also gift vouchers that could be issued following an order made through the Internet channel. There is an opportunity for ATMs to be the physical fulfilment component of marketing campaigns where distribution is important, for instance, redemption of vouchers for prizes or offers.

The fly in the ointment for this proposition is the Internet. Tell me what non-cash functionality cannot be conducted over the Internet that could be done via an ATM? Increasingly, concert tickets and boarding passes complete with bar code information can be issued at home with just a PC and a printer. In this way, the consumer cuts out the intermediary completely and that fulfilment role the ATM could potentially fill is now redundant.

Again, the ATM is a location-specific distribution point and its effectiveness relies on its ability to provide something that is needed at a particular time and place. There is little point in offering an airline boarding pass at an ATM as I can easily print this out at home. On the other hand, if I receive an SMS on my mobile phone saying that I need to redeem a voucher and claim it from a nearby shop within the next hour, then the ATM does become relevant as it can fulfil a specific purpose in a location close to me.

ATM as the cash converter?

In terms of financial-related opportunities for the ATM, there are two key areas—as a top-up for e-wallet applications such as smart-cards, and as conversion to cash for electronic transactions.

E-wallet refills

The promise of the e-wallet, or m-wallet, which we discussed earlier, is best seen in the use of smart-cards such as Visa Prepaid, Octopus (Hong Kong), Ezy-Link (Singapore) and the Oyster (UK). The intention of the cards was to facilitate commercial transactions using a stored value card with no cash involved. Increasingly, we are seeing the emergence of NFC (Near Field Communication) and RFID (Radio Frequency Identification) mobile phones that mimic these smart-cards in operation.

Are there any implications in the use of e-wallets for ATMs or self-service devices? Depending on how the cards are topped up, does the ATM have a role in being the "hub" where money can be transferred across from the bank account onto the card?

In Singapore, the CashCard, which is used in motor vehicles for parking fees and Electronic Road Pricing tolls, can be topped up using bank ATMs. But a majority of the add-value transactions are undertaken at specific terminals placed in logical locations such as car parks and shopping centres where the traffic and the need are, thereby limiting the usefulness of the traditional ATM in performing this transaction.

The other factor limiting the ATM, and for that matter any other physical top-up device, is the use of GIRO, or direct bank account debiting. In this method, the customer's card is automatically refilled by debiting the customer's bank account when it reaches a set balance. This is a popular means of top-up due to its "set and forget" nature.

Cash conversion

While it may be difficult for the ATM to find a relevant role in converting cash to smart-card, it has a lot of potential going the "other way", i.e. converting a remittance from an electronic message into cash.

In the realm of m-commerce and e-wallet, the role of the ATM is crucial as it is the "cash converter". While the hope of m-commerce is that financial transactions can be conducted without the need for cash, without suitable point-of-sale technology, the demand for cash is likely to remain even in developed mature economies. There is no reason to believe that emerging markets will be any different; in fact, they may desire cash more than mature markets.

An example of this need for cash conversion is Globe's GCASH, a Philippine service that relies on SMS to transfer funds between two parties. Apart from people-to-people over the mobile phone network, many overseas remittances are conducted the same way. The cash is given to the agent who then "wires" the funds to the beneficiaries' accounts. The problem there is how to convert the funds into cash. GCASH has a number of ways, including through their network of physical agents, but also through a bank's ATM network. The ATM is the obvious option as the funds can be immediately converted into cash. Services such as GCASH are particularly popular with foreign workers, especially in Asia and the Middle-East, for remitting funds to their families back home.

A similar scheme operates in Kenya. Called M-PESA, it provides for person-to-person fund transfers using mobile phones. Interestingly with this scheme, it is the cash conversion which is a big issue. The shopkeeper who signs up as an agent for M-PESA is required to keep a cash float; this can cause problems in isolated areas and the merchant has to travel to the nearest town to replenish the cash. An obvious opportunity here is for an ATM to be used in the village as the cash converter, resulting in greater security and more convenience. See Chapter 6 for more on M-PESA.

Use of biometrics on self-service

On the technology front, the biggest potential improvement for the ATM is the enhancement of security through the use of biometrics. The most obvious would be using a fingerprint, or facial recognition, to verify the user. There are two advantages in using biometrics.

First, it eliminates fraud. If biometrics are used, it will be more difficult for criminals to observe PIN input and replicate through theft of the cards or through cloning. The second advantage is simply a faster, better user experience. With so many personal numbers and passwords required, the ability to access accounts through a unique biometric feature has an appeal. Biometrics can be added to an ATM very cheaply. NCR estimates it would only add about $120–$200 to the average price of an ATM unit. Given the broad acceptance of fingerprint technology in laptops and mobile phones, customer acceptance would not likely be an issue.

While the biometric hardware is cheap, the effort to modify the software and authentication processes involved requires significant investment at the bank's end. For this reason, currently only two countries have incorporated biometrics in their ATMs—Columbia and Japan. Columbia is obvious because there are security concerns due to the prevailing law and order situation there. Japan is more interesting because security is not as much of a concern; I think that perhaps it's the need to be seen as technologically advanced—essentially a consumer perception—that drives the need for incorporating biometrics in the ATM. In fact, Japan uses a form of contactless biometrics, as Japanese consumers are very hygiene conscious and do not want to be in contact with surfaces touched by other people.

Biometrics has more appeal in emerging markets, especially in poorer, less sophisticated markets where ATMs are a relative novelty. The potential for growth is limited by the need to distribute PINS and retain integrity over the process, i.e. keeping PINS and cards separate, especially in poorer areas where people may not find it easy to remember PINs and the distribution of PINs presents a security risk.

Informational advantage

ATMs could truly start to become an automated "teller" in the next few years. We might integrate Avatar access to customer service and voice recognition, as well as integrate with payment cards, RFID technology and a number of other key technologies that enhance the customer experience.

However, the key improvements will be informational integration tailored to your unique needs. When you visit the ATM, it will inform you of any outstanding bills you need to pay, whether you want to redraw that personal loan you have, whether you want to upgrade that credit limit, or transfer the outstanding balance of your credit card to a cheaper line of credit facility at a better interest rate.

The ATM will offer to deliver messages to your mobile phone relevant to your transactions or your relationship with the bank. The ATM will give you service messages relating to questions you left with the customer service team the last time you called, such as "Your recent request for a credit line extension has been approved."

The ATM will become part of a multichannel customer communication platform. Now with IP-based and multimedia-capable ATM devices, we can do so much more than we are currently doing with our ATMs. The concept of a standard ATM interface needs to change; it needs to be replaced with something that can be populated in real time over the secure IP connection with the bank CRM system.

Advertising and Alternative Use

Interest and credibility in the use of the ATM as an advertising device are rapidly increasing. As we will discuss more in Chapter 13, recent research shows that digital media spend will continue to grow in the coming years. Digital advertising was up 23.3 per cent in 2008, making up more than 10 per cent of all advertising spend globally, and was forecast to grow by a further 22.8 per cent in 2009.

The range of digital options available to media buyers is growing in both size and variety. These range from online/mobile digital opportunities through social networking sites, rich media ads, Bluetooth and SMS, through to outdoor digital media, ranging from retail store billboards and tube displays to in-taxi screens and large format posters—the list goes on.

A few years ago, most bankers would have cringed at the suggestion that their ATMs could be used as billboards, hoardings or media platforms for third-party advertising. Many still feel that way. However, whether you are using your ATMs to advertise only your own products, or using the platforms for advertising products and services of third parties, the potential for the platform has increased significantly in the last few years. Precisely because most banks are generally conservative, those who use these methods of leveraging ATM sites tend to differentiate their brands very quickly.

There are estimated to be more than 1.5 million ATMs globally. In the UK alone, over 130 million different "card holders" will access ATMs annually, and over 21 million access ATMs each week. This translates to 75 per cent of the UK's available cash being distributed through an ATM.[2]

ATM users are a sought-after target demographic for media buyers. The vast majority will have bank accounts and good credit ratings. People feel comfortable using ATMs. They are a trusted customer interface, in most cases used by cardholders at least once a week. In fact, in the US the average usage of ATMs is 10.6 times per month. ATMs may also be the only cost-effective way to engage with certain customer segments that might be difficult for other types of media to reach.

For some banks, ATM advertising is replacing direct mail altogether. Diebold estimates that response rates of up to 20 per cent can be generated on self-service terminals; this is 20 times higher than the average for direct mail campaigns.

The results of any ATM-based advertising campaign can be measured in great detail, usually in real time. According to NCR, ATM advertising is 65 per cent cheaper and 200 per cent more effective than direct mail.

There are three opportunities for presenting an ad or message to consumers at the ATM—when approaching the ATM, during the transaction, and on completion of the transaction.

1 Approaching the ATM
- Vinyls or "skins" that advertise a product or concept.
- Dynamic video advertisement in free play while ATM is inactive.

2 During the transaction (idle machine or processing time)
- Instead of giving the boring "please wait while your cash is delivered" message, or just letting the screen sit

while customers await the next step, present a quick video offer or message for users of the ATM.

○ The transaction time is not lengthened by branded content. Consumer feedback demonstrates that it is preferable for waiting time to be filled with something engaging, rather than a revolving egg timer.

3 On completion

○ A final message after the card is withdrawn.

○ Printed coupon on the receipt. The receipt is transformed into a branded, take-home element of an ATM advertising campaign. It can act as a bar-coded discount voucher, sampling offer, or a proximity prompt for a high street retailer.

Let's simply divide the methods of use into three categories—**external advertising, digital media on the screen** and use of the **receipt media**.

External advertising

The ATM can often provide a platform for a promotion that is under-utilised. For launching new products or services available through the ATM, using the external space on or around the ATM machine itself is a very effective method of reinforcing a campaign. Banks may spend tens of thousands of dollars on external billboards and advertising, when the ATM is right at the point of impact and is a property that they already own; thus there is no external "rental" or media buy element—just the cost of the adware itself.

Ads on ATMs come in different formats, but the two most common implementations are the use of vinyls or so-called "wraps", and ATM "toppers"—plastic housings that sit on top of ATMs to promote the ATM and its access capabilities.

The amusing graphic shown in Figure 7.2 is part of an advertising campaign for a German recruiting agency called Jobsintown. Huge lifelike sticker shots showing workers inside vending machines were used in an attempt to lure people to seek the help of this agency. Although these ads are very clever and humorous, they are only aimed at people who feel they need a change of career.

While this "third-party" advertising generates revenue, the owner of the ATM needs to be assured that there is no conflict with its own brand. The risk being run, especially for a bank-owned ATM, is dilution of its own brand through the association with another company. For third-party ATMs, it becomes a matter of identification of the machine for its core purpose—cash availability. If the customer cannot recognise where the machine is, it then defeats the purpose.

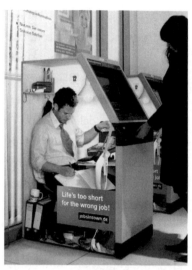

Figure 7.2 Clever "Jobsintown" advertisement on ATM skin
(Credit: Jobsintown.de)

These vinyl or polyurethane wraps range in price from US$200–$3,000 for a full ATM enclosure, to as little as $5 per square foot. Various campaigns in recent times include:

- Visa International promotions for the Beijing Olympics featuring images of athletes.
- Travel or airline promotions integrated with coupon offers for discounts, etc.
- Chase ATM co-branding at Universal Studio Theme parks with an "in your face" cartoon character theme.
- Charity promotions, with an option on-screen to donate for the call to action.
- Confectionary, beverages, snack foods, or FMCG brand promotion on ATMs at grocery stores and supermarkets.
- University campuses branded with school colours, logos or college football team mascots to create loyalty.
- Seasonal themes, such as Halloween and Christmas.
- Landscape or vista themes to differentiate.

We wrap buses, trams and phone booths, so why not ATMs?

On-screen placement and coupons

With the vast majority of ATMs now using the Windows platform and IP-based networking, they are now able to handle more content delivery and advertising functions, and these can be updated a lot more regularly. In effect, anything you can display on your desktop or laptop in a browser you can run on an ATM platform.

The most common implementation of on-screen advertisements are Flash based video or animation, something akin to TVCs in the BANK 1.0 world. Like TVCs, you have a limited time to get a simple message across. So the suggested approach is

Figure 7.3 Even McDonald's are branding their own ATMs

first to have the ATM running a repeating flash video on screen while it is inactive (Figure 7.4). This is interspersed with bank branding.

So something like a 15-second animation advertising BA's latest cheap flights to Paris this summer, then a 5–10 second bank banner display with the logo prominent, followed by the bank's current primary campaign. Once the user inserts his card, the normal transaction process follows. In the case of a cash withdrawal transaction (about 95 per cent of all ATM transactions), instead of the typical message "please wait while we complete the transaction" displayed here while the cash is being retrieved, another short BA promo advertisement plays here. The message about your cash is on its way can still be incorporated. The final step is that the withdrawal receipt prints with a voucher or special offer reinforcing the BA campaign. If you like, this is a mini integrated campaign right there on the single device. Combined with a wrap promoting BA.com or similar, this would be a very unique offering in an increasingly crowded media space. Stage one is to attract the customer to use your ATM.

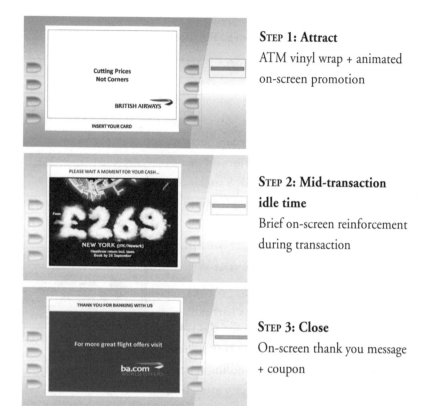

Step 1: Attract
ATM vinyl wrap + animated on-screen promotion

Step 2: Mid-transaction idle time
Brief on-screen reinforcement during transaction

Step 3: Close
On-screen thank you message + coupon

Figure 7.4 Example of third-party advertising on an ATM

Other coupons are presented as vouchers in-store for immediate call-to-action. Research shows that when on-screen advertising is married with a coupon offer in-store, a typical ATM customer spends 20–25 per cent more than a non-ATM customer—that's a compelling reason for retailers to utilise this medium.[3] Given that the placement of an ATM is the second most requested service for retail stores, this makes a great deal of sense for advertisers.

Mobile phone provider "3" ran a proximity campaign, showing ads on ATMs within a certain distance of retail outlets. To drive people to their nearest "3" Mobile store, a bar-coded discount voucher was included at the bottom of every receipt. So with some thought, such advertising can work effectively for advertisers. What are the elements that make ATM advertising attractive to advertisers?

- **Product categories that work.** Frequent ATM advertisers include snack foods, beverages, travel, telecom, fashion and retail.
- **Demographics.** Advertisers can target ATM campaigns, cherry-picking machines by market, zip code and neighbourhood. They can target by ethnic group, income, and even age to some degree, choosing, say, machines on or near campuses to reach college students.
- **Making the buy.** Dispensing coupons with cash at various locations. Typical pricing in the US is $0.08 per coupon and $75 per ATM for 30 days of an on-screen static display. Other groups offer ads on the back of ATM receipts. Pricing ranges from $2,000 for 100,000 impressions to $100,000 for 100 million impressions. Given the dwindling success of direct mail, this looks pretty attractive.

Figure 7.5 Meal upgrade voucher on ATM receipt

Who's advertising on ATM machines? Recent ATM advertisers include British Airways, American Airlines, Tommy Hilfiger, Sargento, Vodafone, "3" Mobile, Volkswagen, AT&T, P&O Cruises, Dr. Pepper, Subway, Pizza Hut and Burger King.

Conclusion

The ATM has to continue to adapt to survive, but in doing so, banks must realise they have very limited time to engage customers who are using ATMs primarily because they are time-poor. Think very carefully about the interface and the engagement method. New technologies integrated into the ATM must improve the speed with which customers transact, and not extend the transaction time frame.

While cash is still king, there are threats coming from mobile devices and smart-cards which will, over time, have a deleterious effect on ATM utilisation. If ATMs become part of the solution, however, then their lifetime will be extended indefinitely. For the more adventurous, this represents new revenue opportunities also. Cash deposit, cheque deposit and passbook updating machines are also an important part of the restructuring of the branch to a more sales and service oriented frontline psyche.

KEY QUESTIONS

The ATM recently turned 40, and it doesn't look like it's diminishing in popularity any time soon. As banks seek to leverage their extensive ATM networks, what are the key drivers for use by customers?

Where should ATMs best be located? When do people use them, and what could prevent them from using a specific ATM? The case of revenue is still a question. Is the ATM a cost centre, or a profit centre? Originally designed to reduce branch load, is it doing its job?

Other self-service devices such as cash deposit and cheque deposit machines are also being deployed. How successful are these? What does the future hold for the humble ATM platform?

The key to self-service is ease and speed of use. Don't get too complicated and when it comes to decisions on what to deploy, strip the process or task down to its simplest form.

Keywords ATM, Cash Machine, Withdrawals, Account Balance, Security, Placement, Shopping Malls and Retail, Usability, Screen Design, Advertising, Wraps, Toppers, On-screen Animation, Coupons, Biometrics, Smart-Card Recharge, Facial Recognition, Fingerprint, Convergence, Travel Insurance

Endnotes

[1]   redi-ATM

[2]   i-Design

[3]   ATMGlobal

8 Navigating Rapid Change Dynamics

It Starts Today

The future success of the world's banks will be inexorably linked to the ways and means they service their customers. Products will mean little in a world where competitors can copy product innovation in hours, technology will be imperceptible to customers who live with constant technological change, traditional advertising methods will be relegated to core branding only, and the only thing that will matter is what we know about our customers and how that makes us a better bank for him, her or them.

The world in which banks emerge over the next 10–15 years will be as significantly different for these institutions as comparing retail banks today with what they were back in 1995. To illustrate the point, in 1995 we didn't have Sarbanes-Oxley or The Patriot Act; most of us hadn't even heard of Anti-Money Laundering (AML) or Know-Your-Customer (KYC); BASEL II didn't exist; you never had to ask the compliance department for approval on a new initiative; you had never heard of credit risk ratings on your customers; there was no such thing as Internet banking (most banks didn't get that until 2000 because it was a fad apparently); there were no IVR-enabled call centres; preferred banking was an idea, not a brand; and a mobile phone was something you carried around in a suitcase (that is, if you could afford to buy one.) A lot has happened in the last 10–15 years.

The Internet now gives many retail banks 50 per cent of their new product applications, self-service and electronic banking accounts for approximately 90 per cent of transactions in the developed world. HNWI clients are by far our most profitable retail customers, and we get more than 50,000 calls from customers to our call centres every day. Compliance, risk mitigation, AML, corporate social responsibility are all huge factors in the corporation today. Regulation is tightening in most markets and we are constantly looking for ways to reduce our operational risk, not just because of the BASEL II accord, but because of the ever present regulator looking over our shoulder. Today's retail bank could not be more different from 1995. If we tried, could we have predicted it? Where will it go next?

Where Will the Changes Impact the Most?

Banks are not yet ready for most of the expected changes that will occur in the next 10 years. The problem is that change is accelerating more rapidly due to **technology adoption diffusion.** So unless banks anticipate those changes, they'll be playing catch-up for the next decade. So where and how are the big changes coming? Pretty much everywhere, but we've classified the change impact into five distinct arenas as follows (Table 8.1):

Table 8.1 Arenas of change and banks' preparedness for them

Arena	Preparedness	Extent of Change	Examples
Platform	We've started	Constant	XML, STP, credit risk automation, mobile, distributed capability, voice recognition, avatars
Channel and distribution	Forget silos, this is our DNA	Severe and rapid	Payments systems, RFID, NFC, automated branches, bank-shops/megastores, contact management
Customer intelligence	We're not talking segmentation	Total re-engineering of customer alignment	Predictive and precognitive selling, behavioural analytics, brand experience
Marketing	Mostly in denial	Irrevocable paradigm shift	The death of direct mail, TVCs, and print; social networks, point-of-impact product placement, Market of One
Metrics	Moderate	Strong move away from financial and branch/product-led metrics	Behavioural analytics, customer profitability, lead to offer ratios, etc

What Are Some of the Trends Dictating Change?

Writer and management consultant Peter Drucker has often been quoted as saying: "Trying to predict the future is like trying to drive down a country road at night with no lights while looking out the back window." The fact is, all that banks can do is to identify trends that are the most likely to impact their business and areas where they are already falling behind. Here are a few areas that warrant serious consideration in forward strategic planning:

Technology Adoption and Disruptive Innovation

We already discussed the rapidly decreasing adoption cycles and the pressure coming to bear due to digitally connected consumers. What will be next? It doesn't matter; what matters is that we need to get faster at utilising and integrating these technologies because our customers already have.

Gridless Customer Experience

The bank's most valuable customers are increasingly mobile and demand secure and instant banking services wherever they are, at any time of the day. HSBC's Global Premier initiative is evidence of this, but in increasing measure customers will be telling their banks where and how they want to connect. But the real issue in an era of unlimited choice will be how to present what's available and help people decide what they want. The information flow we will have to deal with will be immense, so also expect major changes in tools used to process information.

Collective Curation, Social Networking and Web 2.0

YouTube, Twitter, Facebook, Digg, Mashable, Second Life, Flickr, LinkedIn and many more social networking sites and tools have sprung into life and dominance in the last five years. The customers of 2020 will have grown up with these technologies; they are "digital natives". Traditional media for them will be the Web and mobile, while traditional banking for them will be the Internet and widgets. Social networks will be much more efficient at providing trusted advisory services than the bank, and they will compete with our traditional banking services head to head.

Tim O'Reilly, founder and CEO of O'Reilly Media, puts it well: "The essence of Web 2.0 is that networked applications get better through participation. It's true collective intelligence."[1]

Payments and Cash: Chips, RFID, Biometrics, Micropayments, PayPal and AML

We've been talking about the death of cash since the days of Mondex. In Kenya, mobile payments have already overtaken credit cards as the dominant play, and mainstream markets such as Japan and Korea will soon follow. As payments capability become integrated into our phones, watches, clothing and even our body, the ability to transact securely and instantly will push the deposit taking business into historical irrelevance. The death of cash will also solve our anti-money laundering headaches, as every transaction will leave an electronic trail. However, personal and digital security will become even more important.

Pattern Recognition, Predictive and Precognitive Selling, Permission Marketing

Customers no longer read ads that are irrelevant. In the US, the introduction of TiVo, YouTube and iTunes has obliterated the TV commercials market, with Coke and Pepsi cutting TVC budgets by 60–80 per cent over the last five years. Executives are demanding better results than the 0.4–0.5 per cent success rate advertisers get from direct mail campaigns, and direct mail in 2009 is at the lowest level since 2000. Broadcast and mass market marketing are already dying for major brands in the US. Customers want products as solutions unique to them. If we don't anticipate their needs, we lose. If we ask them the same questions we've already asked them 10 times before, we lose. Knowing what our customer needs before they know it will be the competitive edge.

The Roadmap for Change—How Long Have We Got?

If I had asked you back in 1994 or 95 to predict the extent of the changes that were likely by the end of the first decade of the new century, compared with where the bank is at today, could any of us have honestly predicted those changes? Not likely. The Internet, mobile phones and other technologies have impacted banks in ways none of us could have expected or anticipated.

In 1999, the technophiles predicted amazing things for the Internet. Despite the so-called "dot.bomb" stock market dumping in April 2001, our estimates were still generally dramatically short of the mark. For example, we predicted by 2005 Hong Kong's Internet penetration would reach almost 50 per cent of the population; by 2005, 76 per cent of Hong Kongers were online.[2] We predicted by 2011 China would surpass the US in Internet adoption, but as of June 2008 China's 330 million connected users[3] are already starting to make the US's 220 million look a little shabby.

In 2000, we estimated B2B e-commerce revenue globally might reach US$1.7 trillion by 2004; by 2004, those revenues were actually topping out at US$2.37 trillion.[4] We suggested in 2000 that mobile growth in developing economies would be slow because they could not afford expensive handsets; in 2009, 15 of the top 17 mobile service providers operate in just four countries—Brazil, Russia, India and China.[5]

In 2000, Google was still a small private company with a new fangled search engine; PayPal was just starting up in Palo Alto; and Wikipedia (2001), Facebook (2003), YouTube (2005) and Twitter (2006) didn't even exist. Google was listed in 2004, and today their market capitalisation is greater than that of Boeing, Cisco, General Motors, Ford, IBM, Hershey, Nike, American Express, Disney, 3M, Alcatel-Lucent, Juniper Networks, Pfizer, and many more.[6] PayPal is now the most popular online payment system worldwide, beating out Visa and Mastercard. More videos get uploaded on YouTube annually than if ABC, CBS, CNN and NBC combined had been airing content since the 14th century.[7] Wikipedia now has around 2.8 million articles, compared with 66,000 for Encyclopaedia Britannica, a 340-year-old publication.[8]

If this does not illustrate the rate of disruptive change, what will? No bank can possibly believe it is immune to such forces. In the next five to ten years, disruptive change will be a constant and it will have a severe structural impact on the retail bank of today. What will be the changes required as a result of these continuing changes? Take a look at Table 8.2.

Table 8.2 Changes required in the next five to ten years

ARENA	CHANGE/ DEVELOPMENT REQUIREMENT	STRUCTURAL IMPACT
Platform	Customer analytics	Analytics will drive product development and marketing, and we will be channel agnostic. Every channel is equally important, branches are just one.
	Straight-Thru Processing	General insurance, line of credit, loan and credit card applications will happen instantly without human intervention.
	Credit risk management	In line with STP, we need to automate credit risk assessment.
	Contact management	Call centre, Web, email, branch enquiries are all supported through a central integrated system.
Marketing	Individual, integrated and interactive	Offers to customers will be individualised, designed as a journey moving from one channel to the next seamlessly. More than 50 per cent of revenue will be from the Web and Web-enabled phones. Marketing will be split into brand marketing (the current team) and customer dynamics.
	Predictive selling and permission marketing	TVC, print, direct mail will make up 25–30 per cent of the total marketing budget. Rather than the current broadcast/push approach, we will ask and capture customer needs so that we only offer relevant and timely products.
Direct channels	Payments technology	Chips, NFC smart-cards and mobile phones, RFID, e-wallets, etc. will mean much more flexible point-of-sale capability outside of the branch.
	Voice recognition and customer contact centre	"Press 1 for English" will be replaced by spoken commands. The multichannel customer contact centre will become a reality.
	Branch automation	We'll see more automation in the branch as we steer customers away from transactional behaviour at the counter to a sales and service centre. Banks of self-service machines at the front will take transaction load as will non-branch channels.

(cont'd on the next page)

Table 8.2 (cont'd) Changes required in the next five to ten years

ARENA	CHANGE/ DEVELOPMENT REQUIREMENT	STRUCTURAL IMPACT
Direct channels	Distributed banking	Primary branches will have started to consolidate into megastore locations where we educate, inform and sell. Direct sales will take place anywhere customers are with mini-branches, and bank-shops will be the norm.
Metrics	Non-financial metrics	The new measures of financial success will be customer profitability, lead-to-offer ratios and Return on Marketing Investment.
	Real time dashboards	Product managers, channel managers and marketers will see in real time the impact on revenue that campaigns and products are having.

Organisational Changes and Impact

Banks will have to make some significant changes to the way they are organised. Traditionally, the key measure of success on the retail banking front has been revenue or the number of products sold (primarily through the branch). Corporate banking has been physically separated from retail, but likewise has been focused on product revenue. Therefore the business has been split into divisions such as Retail, Corporate, Treasury, etc., and within these divisions the business has been built to reinforce product revenue and sales through the traditional channels. So you get a product focused, branch focused and revenue focused organisational structure—traditionally key factors in a retail bank's success. Marketing continues to be focused on generating branch activity and product acquisition, and banks maintain separate departments for the call centre, the Internet and ATM, IVR and branches.

In the **BANK** 2.0 singularity that banks find themselves in now (and it's only going to get more acute), the realisation is that product and network are no longer differentiators—people and service capability are. The traditional measurement of revenue and product ratios can only make sense in reference to customers and segments today. Traditional branch metrics have become of limited use, because most of the transactions take place outside the branch. Fifty per cent of High Net Worth customers (the most profitable customer group) actually use the internet as their primary channel these days—by choice! So why do banks maintain a structure that is built to promote and build branch activity when patently customers are moving away from this paradigm to a multicontact strategy combining branch and direct channels?

The future of the bank reveals a very different organisation structure. While product teams such as credit cards, wealth management, mortgages, etc. remain pretty much the same, the entire customer-facing organisation and supporting platform need to change radically. The hardest of these changes to swallow will be in the marketing department and the branch distribution structures. IT support becomes more about the fabric of the business, so as a result, we have to split what we'd normally call "IT" into three different supporting functions with different priorities.

Table 8.3 overleaf lists how the organisational shift might take place within the bank. These are very critical components to anticipating and adapting to **BANK** 2.0. In addition, I'll focus somewhat on how the product teams will be affected by all this. This is a factor in the argument of Distribution versus Customer-led philosophy in any case.

From distribution-led to customer-led

Traditionally the branch and the direct sales force have been at the core of everything a retail bank does. It's a cultural thing. It's not so much a branch issue in reality; it is the total dedication to the traditional distribution model. Most banks today don't even classify the Internet and mobile as "channels"; these are normally relegated to alternative channels. It's like the geeky, socially-challenged cousin whom you invite over for family occasions out of respect, but everyone feels a little sorry for him and thinks he's a misfit compared with the rest of the clan. Suddenly, one morning you wake up and find out that the very same cousin is on Bloomberg TV IPO'ing his new start-up and he's being hailed as the next Larry Page or Sergey Brin and you still can't figure out what it's all about.

These so-called "alternative channels" are organisationally separated from branch and "serious" distribution channels because they've often been

Table 8.3 Possible organisational shift within banks in the near future

Traditional Structure	New Structure	Focus
Branch management	Frontline network and partner management	All physical touchpoints, both within the branch network and through third parties, supported by customer dynamics, channel management and strong partner management.
Marketing	Brand marketing	Analogous to the current marketing function, but shifted to new media and limited old media.
	Customer dynamics	An organisational capability to generate timely and relevant offers to smaller and smaller segments, aiming at a market-of-one approach.
Branch distribution, "alternative channels", call centres, Internet banking, ATM, etc	Customer channel management	Combining service oriented architecture, contact centre (voice, IP, email, etc) and behavioural analytics.
Information technology	Content deployment	Content publication supported by customer dynamics.
	Customer dynamics	Managing customer intelligence through customer data mart and customer analytics.
	Customer channel management	Managing contact centre and supporting channels.
	IT	Maintaining systems backbone and core technologies.
Transaction services	Enablement	From middleware to payment systems to settlement systems, transactional technology will be all about customer fulfilment as fast and accurately as possible.
Treasury and cash management	Transactional services	Treasury will be largely automated by technology, especially as STP kicks in.
Strategy groups	Innovation	A multidiscipline (product, IT, marketing) team that stimulates innovation and manages new programme initiatives.

considered IT playthings, rather than core business essentials. Even today, with the Internet for many banks representing more than 50 per cent of their transaction load and 30–50 per cent of new product revenue, there is almost incredulity that this is possible. More often than not, there is an undercurrent of talk that such results must simply be statistical anomalies, or worse, such results are ignored because they don't fit the expected model. If you are a banker and you feel I'm being too harsh, let me ask you one simple question: **In the organisation chart, where does your Internet channel line manager sit in relationship to your distribution head?**

In **BANK** 2.0, the primary shift is a move away from a physical "distribution channel" psyche to a total customer channel mentality. This move requires a rethink on a number of levels. First of all, customer revenue and profitability drive to the top of the metrics scorecard. Revenue through the branch channel becomes a supporting metric. More critical for branches is the net contribution to the cross-sell and up-sell targets that generate customer profitability. A monthly branch revenue figure which bears no relationship to customer measures as a core strategic driver is simply an outcome of a strategy, it is *not* the strategy.

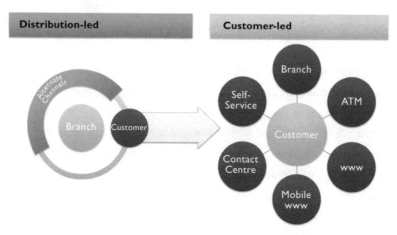

Figure 8.1 The primary culture shift for **BANK** 2.0

Whereas in the BANK 1.0 days we would be engineering the latest product to be offered through the branch network, in the **BANK** 2.0 paradigm the product team has to think a little more laterally about positioning products across a range of channel options, rather than attempting to retrofit products designed for over-the-counter or direct sales force offering to a digital dialogue.

There are sound regulatory reasons for this as well. Separating counter-party risk from operational risk is an inherent principle in the Basel II framework, and as such, a clear separation between the manufacturing and origination of a product versus the distribution of the product is a key feature of these global risk standards.

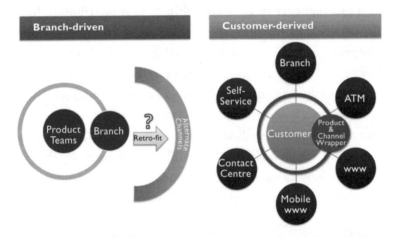

Figure 8.2 Product origination needs to be channel agnostic

In most of the banks I've worked with, apart from the odd "e-saver" account or similar, the creation of a new product takes part totally devoid of any thought on how the product will be distributed through non-traditional channels. Some products retrofit very easily onto digital channels, for example, time deposits. Generally, the fewer features a product has, the easier it is to multifit it across different sales conduits. To be fair, many banks already customise products based on segmentation. We see credit cards and savings accounts customised based on net worth, credit cards for the female demographic (probably due to their propensity for shopping), student loans, general insurance products for families, and many more. Segmentation is a key input into the design of these products, but channel rarely factors in during the design phase unless the product is perceived as a channel specific solution.

Product teams need to move to a matrix approach to product generation and design, where channel is an input, or create products that are channel agnostic. Channel specific artefacts are created after the core product is packaged, and these become **wrappers** that are aimed at a specific channel or segment/channel combination. For example:

- Physical application forms are replaced by a **list of fields or data elements that are mandatory only;** physical application forms or electronic entry forms are a channel specific wrapper.
- **Compliance, credit risk and/or STP parameters** are a core deliverable of the product team and supporting departments. However, compliance needs to think multichannel, for example, supplying three months' worth of bank statements over a mobile phone app form is never going to happen.
- **Benefits to the end user** are a core deliverable, not features. These are correlated by segment.
- The customer dynamics team needs to generate a list of **cross-sell and up-sell drivers** for the specific segment, so that specific campaign instances (customer offers) can be generated at the wrapper stage.
- Creative design needs to start with two core outcomes— **message** and **call-to-action.** Concepts generally need to be done for at least three core media types—print, Web and mobile (MMS)—for each as part of each creative brief. Most current media is *not actionable* and won't fit in the new customer-led dynamic.
- In the case of a multisegment product, mapping offer to segment at the wrapper stage would produce the **media buy triggers,** rather than media buy being solely campaign based.

So the product is derived for a customer segment or segments, and it is designed to be packaged for each channel with a wrapper that amounts to either an acquisition attempt, or a cross-sell or up-sell offer. Either in the instance of a new customer or for an existing customer, the outcome should be a call-to-action that facilitates the sale, not a brand or product message. Let's take a specific example to illustrate. A good example might be a personal loan offer at a specific time of year, such as a tax loan at tax time, or a travel loan leading up to summer vacations, or a student loan around enrolment time. Here is an example of how it might map out:

Personal Loan
- Loan Amount available (eg. 3 × monthly salary)
- Underlying Interest Rate (eg. 6.25% or 4% above LIBOR)
- Duration or Term of Loan (eg. 3 months min, 24 months max)
- Credit Risk Rating limit: Good to Excellent (ratings driven from internal credit risk system)
- Core data requirements
  - All Customers: Amount, Term, Preferred or Normal Rate (internal)
  - New Customer
    - Name, Contact Details Set, Employer Details Set
    - Annual Salary, current credit card balance
    - Bank contact if not our bank

Target Segments
- Customers with salary account and 12 months' history
- Customers with good to excellent credit risk rating
- Customers with maturing personal loan (redraw offer)
- New target, aged 25–45 working professionals
- New target, expat professionals as possible preferred segment up-sell
- Inactive customers with deposits in excess of US$25,000

Benefits
- Existing customers: Pre-Approved or Approval in 60 seconds
- Funds the same day
- Choose the term that suits your budget
- Competitive rate

- Offer 0.5% privilege rate for preferred customers for 30 days
- No penalties for early pay out
- Automatic redraw facility for existing HSBC customers
- Themes: Renovate Your Home, Upgrade Your Car, etc.

Table 8.4 presents a fairly vanilla product offering with some interesting execution options for either customer acquisition or cross-sell. Obviously, the wrappers could be further segmented into different groups of customers or different themes. We haven't touched on traditional print media as a promotion element here, but you could use this for **drive to Web** or **drive to branch** also. Clearly, however, the approach of using a core product offering and "wrapping" the product for use within an appropriate channel gives a lot more flexibility than trying to retrofit a branch product after the fact.

Table 8.4 Products and their wrappers by channel (customer dynamics team)

	WRAPPERS BY CHANNEL
Public website (new and existing customers)	• Landing page, possible dedicated URL (e.g, getaloan.com.sg, newcarnow.com) for link with print campaign • Third-party "Apply Now" rich media banners • Link to Internet banking process for existing customers
Internet banking (existing customers)	• "You have been pre-approved" banner and secure message • "Pre-approved credit options" contextual hyperlink on account summary page • "Transfer Credit" hyperlink on credit card summary • Add new listing under Account "Personal Loan" with maximum pre-approved amount, i.e. $50,000 Pre-Approved • Add "pre-approved credit options" to left-hand navigation
ATM	• Rich media ad playing during idle time • "You have been pre-approved" message for existing customers on close of transaction • Call back request option • Promotion on coupon for selected segments: "It takes less time to get a loan approved than it took to get your cash."
SMS/MMS	• Personalised message with Web app link: "Mr. King, HSBC has pre-approved you for a personal loan. Are you interested?"
Mobile app	• New button for "Pre-approved Credit" • For existing customers, just two fields—term, amount (pre-set drop down list for options) • Call back with approval and compliance procedure notice

(cont'd on the next page)

Table 8.4 (cont'd) Products and their wrappers by channel (customer dynamics team)

	WRAPPERS BY CHANNEL
Branch	• Banner stands as per themes, TV and poster board promotions • Pop-up cross-sell on branch dashboard • Focus on preferred customers for month of January, offer discount interest rate • Digital application—green, i.e. no paper? Can we sign on digipad? (check with IT and branch services)
Other?	• See if we can integrate personal loan as a payment option on CathayPacific.com and BA.com, instead of credit card for existing customers (does cookie technology allow this level of granularity?)

I know many banks have already been looking at options for deploying product over new channels, but generally the banks start with the standard branch or physical distribution options first. In the new **BANK** 2.0 world, retail banks need to be agnostic as to which channel is used first or most prominently for the product promotion and roll-out. The wrapper approach allows much more specific channel metrics for each product and customer segment as well, garnering much more constructive data than is currently captured. Basically we'll know who, when and via what channel the customer committed to a product—rather than assuming the branch is always the first place they will look.

Information technology to customer data management

The primary IT shift in **BANK** 2.0 is that data, systems, platform and applications all need to be dedicated to the following key pillars:

- Understanding customer behaviour.
- Anticipating customer actions and needs.
- Enabling rapid customer fulfilment.
- Better service architecture.
- Channel agnostic content support.
- MI (Management Information) integration of customer analytics.

IT has long been held as the "functional" enabler of the bank. Many a project, product or initiative has been scuttled due to lack of IT resources, time or budget constraints. In the new **BANK 2.0** universe, with technology and interaction being such a core element of the bank's ability to deliver, the role of systems has to evolve to one of being part of the bank's DNA or very fabric of existence. Now this doesn't mean that we simply expand the IT budget and make a bigger IT empire. In the new operational paradigm, we have to create multiple "system" functional areas that support the bank in its mission. Traditionally these would have been considered IT, but in our new view of the world these will become customer enablement platforms.

To date, most bank internal reporting systems have been based on one of three elements of the business, namely **financial metrics** (profit/revenue generated, costs accrued), **customer metrics** and **service metrics.** The technology platform has been bundled into simply an "IT" function, requiring IT to support or sign off on practically any piece of software or hardware in use. With increasing deployment of multichannel solutions and the need for much more customer engagement via technology, some of these boundaries need to change.

The three key informational departments emerging in the new **BANK 2.0** world will be **Enablement** (the closest analogous to the current IT department), **Customer Dynamics** and **Content Deployment.**

The driving force within this space will be customer analytics. While today considered an IT function—that is, I have to talk to the head of IT if I am a customer analytics vendor—the fact is that customer analytics and resultant marketing strategy need to be, right now and today, the most important management information system in the bank. This is a dramatically misunderstood space.

The customer dynamics function

Customer dynamics is not a product team generating new product ideas for niche segments. **Customer dynamics becomes the interface for every non-brand related sales message or offer coming from the institution to the customer.** Staff from marketing, IT and from each product team are interspersed into the team, as well as segmentation specialists and data analysts to review and process analytics data. The shift from broadcast, shotgun advertising of product offers, to targeted, laser-focus market-of-one messages and dialogue is the NOW of bank marketing. I say the now because it is not the "future"—it is where banks should already be today, but they are not.

Traditional marketing departments lack the skills of innovating customer messages in the **BANK** 2.0 environment. Universities still teach Marketing 101 to this day based on the reinforcement of core brand features or triggers and the ability of a marketing message, sometimes called an advertisement, to trigger brand recall at the time you select your purchase. This is an outdated, outmoded relic of 20th-century marketing doctrine. The 21st century is all about understanding the customer and producing the right offer, at the right time, for the right customer.

To affect that capability, we need to start with a basic understanding of our customers. If you are a banker, can you tell me which products within your retail repertoire of products best fit specific channels? The likelihood is if you ring the head of the Internet or the contact centre today, they wouldn't even have this information readily available for one channel, let alone every channel in the bank. This drives **new customer analytics system** requirements.

The customer dynamics team needs to be the owner of all customer marketing in the **BANK** 2.0 paradigm. This means they need to select core segments of customers based on behavioural analytics, then they need to craft multiple up-sell or cross-sell offers for seeding through each channel. For specific product deployments, the customer dynamics team will liaise with the product and marketing teams below, but the product and marketing teams do not deploy campaigns without the customer dynamics team's approval. The customer dynamics team will also look after the critical element of **point-of-impact** technologies and preparing or cacheing the appropriate sales messages.

The content deployment function
In an increasingly aware marketplace, content is key to educating and informing customers. **The content deployment team becomes the**

manager of every element of content, information or data deployed to customers on a regular basis that is a service or informational message. They also manage the evolution of user interactions and systems interface redesign, with their embedded usability and information architecture specialists. This team could also be called the customer interfacing team, but I don't want to make it sound like a direct sales or contact centre function.

The amount of content that is deployed to customers is increasing, as is the number of outlets we have to reach those customers, including Web 2.0 social networking sites, new devices such as Google's NexusOne or the iPhone and other such channels. Controlling the quality and clarity of information that reaches the customer is the sole domain of the content deployment team. The content deployment team works most closely with the channel management team to ensure that the systems and publication mechanisms for distributing content and messages to customers are robust and accessible for key content authors or systems within the bank.

Let's not confuse this with the content management systems of the late 90s as the dot.com era took hold. This is about controlling the quality, frequency and appropriateness of information and service messages that we impart to customers.

The role of IT in risk and compliance

Significant regulatory changes and compliance requirements have been introduced in recent times. There have been different factors promoting this change, including the collapse of Enron, WorldCom, Tyco and then 9/11 in the United States, and in 2008 the collapse of major financial institutions such as Bear Stearns and Lehman Brothers. The 9/11 and 7/7 (London bombings of 7 July 2005) shocks generated considerable pressure on anti-money laundering disclosure agreements as an attempt to halt or reduce terrorist funding.

In order to support these growing requirements adequately, IT capability in risk management, compliance and reporting will be absolutely essential. More important will be the need to simplify the processes behind this capability so that customers are not negatively impacted. Some have commented that KYC means "Kill Your Customer" (with compliance). Indeed, in the mid-90s, we may have asked the customer to complete or review three to five pages before we would open an account, but in 2009 that has increased to 18–20 pages.

Compliance needs to understand the negative risk of increasing workload on the frontline in respect of customer service perception. They need to start thinking about their function as an **enabler** of core business with customers, rather than just risk mitigation. They need to be lobbying regulators to help them adapt and make their processes more user-friendly, while retaining security of identity and the assets of the customer. IT needs to lead the charge in improving systems to achieve this.

Compliance and IT departments need to work as consultants with the customer dynamics team to re-engineer the way we do our compliance checks, credit risk assessment and product application processes.

Conclusion—Tradition or Innovation?

The changes are going to be rapid and significant. The good news is that initial development of channel support and some of the platform capabilities are already in place. The complication, however, is that since the early 1980s banks have been comfortable in retail growth that has generally been leveraged off just three core retail segments—mass market, mass affluent and High Net Worth customers. As a result, the business has been built around products to support these segments and offerings through the primary retail channel—the branch.

There used to be a story running around the banking fraternity to explain how banking worked. The saying went that you took deposits at 3 per cent, lent them back out at 6 per cent, and you make ... it on to the golf course by 3pm. Banking used to be a pretty simple game, but margins and the mechanics of banking have flipped banking on its head. It used to be that you could not lend out more money than you had in deposits, but banks are lending out more and more money through increasingly complex debt apparatus that are invariably also more difficult to find on the balance sheet. Central banks and capital adequacy requirements have also been getting more and more relaxed over time, despite the impression that regulation is getting tougher.

Metrics for success are going to have to change rapidly in the **BANK** 2.0 world to encourage the requisite maturation in marketing, IT and other business competencies to support a much more dynamic and fluid operation, with much greater granularity in customer groupings. Indeed, traditional measures will have to take a back-seat as segmentation rapidly fragments into smaller collectives and as product revenue shifts almost entirely towards direct channels.

Profitability will also be adversely affected unless these changes take place. Today, margins are being squeezed on all low-involvement products. Many retail banks can't even make money on mortgage products in the first three to five years. With loss of differentiation on product, the ready availability of comparative sources and non-bank offerings, margins are going to be further squeezed. When banks ramp up rates without a good reason, customers will revolt; they will blog and tweet, and banks will lose out. Add to this the increasing cost of labour and property to maintain branch presence, higher marketing costs to support drive-to-branch, and the lack of "service" perception because other direct channels lag the branch in respect to brand experience—and we have an emerging crunch in profits and customer retention that will likely cripple future growth and market credibility.

In the end, a bank is just two things—an organisation that provides service for customers (in turn generating revenue), and a company that provides returns (through dividends and capital gain) for shareholders. Today, banks have largely forgotten about customers because they maintain that they **own the infrastructure** that the financial system runs on. But that is no excuse. To provide the level of service that customers have come to measure banks on, retail banks will have to break out of the traditional product and revenue supply chain that has grown through traditional branch banking, and reinvent the business around customers and the way they work with the bank. Why? Because customers will choose alternative mechanisms increasingly for many of the "functions" that used to be considered the sole domain of banks, but are now being provided by third parties, such as PayPal and Mobile P2P payments.

Direct channels will dominate in the **BANK** 2.0 world not only because customers will use these channels the most, but also because from a business point of view, margins can only be salvaged by lowering the cost of delivery. Marketing will increasingly have to justify their existence in terms of customer profitability and ratio of eyeballs to product/revenue conversion, so they will have to understand customer behaviour intimately rather than rely on broadcast methods of advertising which no longer work adequately.

Offers will be a result of better customer analytics. Core IT and marketing teams will have to support much more aggressive and broad content creation and ownership, as this will become the lifeblood of engaging the customer. Leadership in the bank will be chosen because of a true passion for the customer offering, and they will be channel experts, agnostic to branch and traditional marketing approaches. The skill sets of the distribution and marketing teams will be turned on their head over the next five years. It has to change—dramatically and permanently.

The future, in many ways, has already begun. The only question remaining is how will your bank make the journey? One thing is for certain—they better get started now.

Endnotes

[1] radar.oreilly.com – Tim O'Reilley's blog

[2] Internet WorldStats. http://www.internetworldstats.com/asia/hk.htm

[3] "China's Internet population hits 333 million", China Realtime Report in *The Wall Street Journal*, 17 July 2009

[4] E-Marketer, Forrester Research

[5] "Global Mobile Subscriber Growth Slows", by Gary Kim. Mobile Unified Communications, 4 September 2009

[6] Various sources: NASDAQ; DJI market capitalisation figures; finance.google.com; reuters.com

[7] Various sources: YouTube.com; ABC.com; CBS.com; CNN.com; NBC.com

[8] Sources: Wikipedia.com; Encyclopaedia Brittanica

Part 03

The Road Ahead— Beyond Channel

9 Deep Impact—Technology and Disruptive Innovation

YOU'VE undoubtedly heard of "Silicon Valley". Did you know why it is called Silicon Valley? You might think it is because all the dot.com 2.0 companies inhabit this region of California. But you'd be wrong. We have to go much further back to the 1950s to find the origin of the term. It must have something to do with computer chips because microchips are made of silicon.

Well, in 1947 a gentleman by the name of William Shockley, along with John Bardeen and Walter Brattain, invented the transistor. For this innovation, the three were awarded the Nobel Prize in Physics in 1956. The attempts of Shockley to commercialise the transistor led to the formation of a group of companies in California specialising in the manufacturing of these components. During the 1950s and 60s, there was a great deal of speculation in the markets about "tronics" or the ability to capitalise on these "new" technologies and advances. In fact, there was a stock market bubble in the 60s based on the so-called "tronics" hype.

On 19 April 1965, Gordon Moore, the co-founder of Intel Corporation, published an article in *Electronics Magazine* entitled "Cramming more components onto Integrated Circuits". In that article he stated a law on computing power that has remained consistent for more than 40 years, a law that drives technology development today and for the near future.

> "The complexity for minimum component costs has increased at a rate of roughly a factor of two per year ... Certainly over the short term this rate can be expected to continue, if not to increase. Over the longer term, the rate of increase is a bit more uncertain, although

there is no reason to believe it will not remain nearly constant for at least 10 years. That means by 1975, the number of components per integrated circuit for minimum cost will be 65,000. I believe that such a large circuit can be built on a single wafer."

Gordon Moore's prediction, *Electronics Magazine*, 1965

Moore's Law

The term "Moore's Law" was reportedly coined in 1970 by the CalTech professor and VLSI pioneer, Calvin Mead.[1] Essentially what this meant was that Moore predicted **computing power would double every two years.** Since 1965, that law has held true and remains the backbone of classical computing platform development. But what all this means is that since 1965, we have been able to predict reliably both the reduction in costs and the improvements in computing capability of microchips, and those predictions have held true.

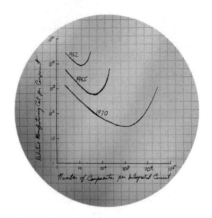

Figure 9.1 Gordon Moore's original graph predicting transistor growth
(Credit: Intel Corp)

In reality what does this mean? Let's put it in perspective. In 1965, the amount of transistors that fitted on an integrated circuit could be counted in tens. In 1971, Intel introduced the 4004 Microprocessor with 2,300 transistors. In 1978, when Intel introduced the 8086 Microprocessor, the IBM PC was effectively born (the first IBM PC used the 8088 chip)—this chip had 29,000 transistors. In 2006, Intel's Itanium 2 processor carried 1.7 billion transistors. In the next two years, we'll have chips with over 10 billion transistors. What does this mean? Transistors are now so small that millions of them could fit on the head of a pin. While all this was happening, the cost of these transistors was also exponentially falling, as per Moore's prediction.

In real terms, this means that a mainframe computer of the 1970s that cost over $1 million had less computing power than your iPhone has today.

It means that the USB memory stick you carry around with you in your pocket would have taken a room full of hard disk platters in the 70s. Have you watched the movie Apollo 13? Remember they were trying to work out how to fire up the Apollo Guidance Computer without breaking their remaining power allowance? Well, that computer, which was at the height of computing technology in the 70s, had around 32k of memory and ran at a clock speed of 1.024 MHz. When the IBM PC XT launched in 1981, it was already about eight times faster than the Apollo computer.

Figure 9.2 The Apollo Guidance Computer, circa 1970
(Credit: ComputerHistory.org)

The next generation of smartphone we will be using in the next two to three years will have 1 GHz processor chips. That is roughly one million times faster than the Apollo Guidance Computer.

Theoretically, Moore's Law will run out of steam somewhere in the not too distant future. There are a number of possible reasons for this. Firstly, the ability of a microprocessor silicon-etched "track" or circuit to carry an electrical charge has a theoretical limit. At some point when these circuits get physically too small and can no longer carry a charge or the electrical charge "bleeds", then we'll have a design limitation problem.

Secondly, as successive generations of chip technology are developed, manufacturing costs increase. In fact recently, Gordon Moore said that each new generation of chips requires a doubling in cost of the manufacturing facility as tolerances become tighter. At some point, it will theoretically become too costly to develop the manufacturing plants that produce these chips.

Lastly, the power requirements of chips are also increasing. More power = more heat = bigger batteries, etc. At some point, it becomes increasingly difficult to power these chips while putting them on smaller platforms. Even today, some laptops are literally getting too hot to put on your "lap".

So when will this all happen? Gordon Moore himself predicted the end of Moore's Law around 2022. Other predictions range from 2015 to 2060. But industry analysts have "called" the end of Moore's Law more than a few times before. Additional transistor densities may be achieved in other ways too, such as via 3D semiconductor structures, or more exotic approaches such as carbon nanotubes, silicon nanowires, molecular crossbars and spintronics.

Even if Moore's Law does end, this does not mean that technology then stagnates and hits some permanent performance ceiling. If transistor size becomes a constant, then the burden of computer development will simply shift to other parts of the classical computing ecosystem. There are plenty of opportunities for improving computing performance in other areas, such as design of the device or system itself, software and firmware efficiency, new board and network architectures that do more work with less silicon.

The fact is that systems such as Windows and much of the software we have today could be far more efficient, but they remain bloated with legacy code that has been kept over successive generations of development.

The usable limit for semiconductor process technology will be reached when chip process geometries shrink to be smaller than 20 nanometers (nm) to 18nm nodes. At those nodes, the industry will start getting to the point where semiconductor manufacturing tools are too expensive to depreciate with volume production, i.e., their costs will be so high that the value of their lifetime productivity can never justify it. Len Jelinek, Chief Analyst, iSuppli, June 2009

How this translates is that for the next 10–15 years (a very long time in business life-cycles), if you are in banking, you will remain under constant pressure to integrate better, faster technologies and techniques to stay relevant to customers and to stay competitive. If you aren't constantly updating your technologies, platform and multichannel capabilities within the bank, you WILL fall behind. This is just a fact of life—get used to it.

After Moore's Law

When we look further into the future, there are really only two promising solutions that will replace the silicon paradigm that underlies the flawless performance of Moore's Law to date. Those two solutions are quantum computing and DNA (or biological) computing.

Quantum computing essentially utilises the quantum state of qubit (the equivalent of a normal bit/bite in computing terms, but at the quantum level). Like a traditional bit, a qubit has an on and off state, but whereas a bit can only be 1 or 0, a qubit can also produce a superposition of both states. Thus, depending on configurations, implementation, the principles of entanglement and superposition (quantum mechanical phenomena), a quantum computer will likely operate on an underlying bit structure that contains at least eight different three-bit strings. But because of the nature of quantum mechanics, it can simulate the calculations of almost any combination of results simultaneously.

This means a completely different type of programming would be required, but it results in massive computing power. Programmes, calculations or simulations that would take weeks, months or even years to complete on today's platforms could be executed in real time almost instantly. Chips the size of a grain of rice would be more powerful than today's supercomputers, and use almost no power at all.

The other promising replacement for silicon technology is **DNA computing** which uses DNA, biochemistry and molecular biology. It was first demonstrated as a concept by Leonard Adleman of the SoCal (University of Southern California) in 1994. Adleman demonstrated a proof-of-concept use of DNA as a form of computation which solved the seven-point Hamiltonian path problem. He used an oligonucleotide, which is just a really fancy name for a polymer. But if you've ever watched an episode of CSI where they take a piece of evidence with a suspect's DNA and put it in a solution to identify to whom it belongs, then you are watching one typical use of oligonucleotides, as they are often used to amplify DNA in what is called a polymerase chain reaction.

Ok, ok, enough of the technobabble ... well almost.

Materials Science

The A380 Airbus and the Boeing 787 aircraft are unique as commercial aircraft go because 20 years ago we simply could not have built these aircraft with the materials available at the time. We're talking in the 90s—not exactly the Dark Ages. However, in the last 20 years, some remarkable advances have begun to appear in the way we can manipulate and create materials for use in construction, in devices, and even in genetics and medical science.

What these new materials enable us to do is to think of almost any commercial application of technology and work towards it in the next five to ten years. We are literally getting to the point where we could build a mobile phone that is effectively too small to use. We are getting to the point where notebook computers are so powerful that I could soon duplicate the entire functionality of a small bank on something I carry around in my backpack, if not for the data storage requirements.

Let's discuss briefly three of the key areas where materials science is going to affect retail banking, or consumers of retail banking, in the short to medium term:

1. Screen and "touch" technology.
2. Electronic paper and flexible circuit boards.
3. Battery technology (we'll discuss this later in the section "All this data, all this power" on pages 295–97).

Screen and "touch" technology

Have you ever tried using your laptop out in the direct sunlight? It sucks, right? Traditional LCD and plasma screens do not cope effectively with this environment very well. Over the last couple of years, a few companies have been introducing OLED or Organic LED screens into their mobile devices, TVs and in laptops currently in development.

The advantage of OLED technology is a super bright, extremely thin screen that takes almost no power to run, but retains the resolution, colours and capability of traditional LCDs and plasma screens—in fact, much better. Within a few years, we'll have wallpaper-thin 60-inch TV screens that take less power to run than your average light bulb. In fact, these devices run off 40–50 per cent less power than the most energy efficient LCDs currently in the market. With contrast ratios of a million to one, that is pretty impressive.

Sony and Samsung have thus far led the way in the commercialisation of OLED technology. Recent releases of OLED TVs, such as the Sony XE-1, provide a flat screen TV with a screen that is just 3mm thick. In April 2008, Sony demonstrated a 3.5-inch display screen just 0.2mm thick that displayed a resolution of 320×200. They also recently demonstrated a 21-inch flat screen TV that was only 1.4mm thick. Samsung, not to be outdone, has recently demonstrated a mobile phone that has a foldable display screen, manufactured using OLED technologies.

Samsung and Sony are both working on laptops that could very soon make their way into production. The Sony VAIO "Contrast" concept is such a device.[2] The Contrast VAIO laptop uses a foldable seamless OLED for the display and the keyboard, but as the videos show the keyboard can fade away and the whole thing can display something else. This concept has no restrictions on layout and size, and is extremely durable and shock resistant. It is made of high performance flexible bioplastic, and it would be thin enough to slide under a closed door if it went into production. Sony has also prototyped a range of "contrast" concept devices based on OLED and bioplastic polymer technologies. A wearable Walkman that can be worn as a bracelet is one such concept. The near future of this technology is called QD-OLED or quantum dot OLED, and could be even thinner and sexier than the current OLED standards.

3D technologies are also rumoured to make an appearance in the near future, with LG, Toshiba and others working on various technologies to represent 3D images without the use of those very attractive red and blue cellophane glasses.

Electronic paper and flexible circuit boards

Very closely related to soft-screen or OLED improvements is the area of electronic paper. This has been termed the most significant development in print technology since the invention of the printing press by Johannes Gutenberg in 1440.

Figure 9.3 (top left) A flexible OLED TFT display; Figure 9.4 (top right) Sony VAIO concept with OLED/Bioplastic technology (Credits: SONY)

It was actually in the 1970s that Nick Sheridon at the Xerox Palo-Alto Research Centre developed the first electronic paper. This e-paper, called Gyricon,[3] consisted of polyethylene spheres embedded in a transparent silicon sheet. Depending on whether a negative or positive charge is applied, the spheres would translate into a pixel that emits either a black or white appearance, thus looking a lot like normal paper.

The Amazon Kindle and Barnes & Noble Nook are both examples of implementations of e-ink technology. They implement a type of technology known as electrophoretic display. This is essentially an information display that forms images by rearranging charged pigment particles using an applied electric field. The Kindle and Nook devices use a hi-res active matrix display constructed from electrophoretic imaging film manufactured by E Ink Corporation.

So the big question is, when will e-paper be ubiquitous and when will Rupert Murdoch stop printing newspapers? Also, is e-paper as "green" as the current paper we use in our daily lives, or is the carbon cost to produce e-paper going to have a negative effect?

Already the Kindle, Nook and other devices are making their impact on the popular psyche. However, in the next five years, depending on battery developments, we would expect real e-paper to be deployed that is not much thicker than an existing wad of papers. The e-paper would be highly portable, flexible enough to roll up, give you a modest four to six hours of battery life, and be able to simulate newsprint, books and

Web pages in either grey-scale or colour. It would download "books" or newspapers using wireless. However, this would be a device for reading print, and would not be capable of full video. Thus, it will be less like a laptop screen and more like "paper", making it a strong replacement for books and newspapers.

The "IN" device—short for "Innovation Newspaper"—is a proposed kind of newspaper that combines a flexible e-paper display and WiFi technology. Place the device in its cradle before bed, set the embedded alarm, and wake up to an updated feed of news for your morning coffee. I'm sure you'll agree that this type of innovation will be a real threat to the traditional broadsheet. Within five years such devices should be relatively commonplace.

HOT TIP: Tackling the OTP (One Time Password) Two-Factor Authentication problem with e-paper

Two-factor authentication is increasingly becoming a mandatory requirement for Internet banking access and for mobile or phone banking-based third-party transfer requests, obviously the result of increased incidents of fraud, phishing attempts and so forth.

E-paper may present a more elegant solution than the current suite of digital tokens or key fob tokens that are out there. The world's first ISO compliant smart-card with an embedded display was developed by SmartDisplayer in 2005 using SiPix Imaging's electronic paper. This can easily be carried in a purse or wallet.

Figure 9.5 SmartDisplayer's smart-card with e-paper display, combining OTP/RFID capability
(Credit: SmartDisplayer.com)

Figure 9.6 2009 IDEA
BusinessWeek Award
(Bronze) Design by
Seon-Keun Park, Byung-
Min Woo, Samsung
Art & Design Institute
(Credit: Seon-Keun Park,
Byung-Min Woo, Samsung
Art & Design Institute)

All This Data, All This Power?

With all these amazing applications of technology and the increasing pervasiveness of devices that will seamlessly connect to the Internet, talk to one another and incorporate themselves into our daily lives, the obvious question is how will we power all of this? Green energy is increasingly a platform for public discussion. Recent information has shown that while a hybrid car such as a Toyota Prius is an amazingly fuel efficient car, the environmental cost of producing the batteries that power the car almost negates the carbon benefit of the fuel efficiency of the car itself.

In respect of powering devices that are going be a part of our future, there are four separate issues:

- generation of and access to "green" energy;
- portable self-sufficiency;
- longer battery life and more efficient energy utilisation; and
- powering micro devices.

While this is not a book on green technology, there are four areas where we will see significant development over the coming decade from a retail energy perspective.

Fuel-cell technology is still probably the most promising technology for powering cars, homes and enterprises in the near future. A fuel cell is an electrochemical device that combines hydrogen and oxygen to produce electricity, with water and heat as its by-product. Conversion of the fuel to energy takes place not via combustion, but via an electrochemical process, making the process clean, quiet and highly efficient—300 per cent more efficient than burning fuel. Fuel cells operate silently, so they reduce noise pollution as well as air pollution, and the waste heat from a fuel cell can be used to provide hot water or space heating for a home or office. NASA has used fuel-cell technologies in its space programme since the 1960s.

The **battery** in your laptop now has an average of some four or five hours of use in high power mode (i.e. watching a DVD), up from some 15 seconds just a few years ago. Battery makers are challenged to make more significant improvements in this space right now, and batteries now make up one of the most expensive components in your devices. In fact, in an average electric car, it is estimated that the cost of the LiOn (Lithium-Ion) battery means the average production cost of an electric versus normal combustion vehicle is some US$18,000 more. We need a new generation of batteries to cope with all the portable power requirements we are going to have over the next 10 years. As yet, however, such a leap of technology is not on the immediate horizon, but there are some organic and nanotech solutions that are getting closer to commercialisation. In 2009, MIT researchers showed that by incorporating carbon nanotubes and using genetically engineered viruses, they could enhance Lithium-Ion battery capacity, increase conductivity and reduce charge time.

Batteries will also have to get smaller and more flexible. The app phones and smart devices we carry will be built into flexible bioplastics with soft screen technology—devices where a rigid battery form will no longer work. Perhaps our clothes will convert our energy from walking into power that recharges our mobile phone. We have already seen hiking jackets incorporating solar cells for recharging iPods on the go. It's not hard to see where such technologies might lead in the future.

As green energy becomes more affordable, it is likely that we'll see a lot more personal use of green technologies. Solar cells, solar hot water heaters and mini-wind turbines are popular technologies for use at home even today. As we get our electric cars and such mini-generation capabilities, retail electricity companies will pay us back through **Reverse Energy Contracts** which will allow for excess energy produced by our solar cells to be pumped back into the grid, or if our car is fully charged and there is heavy demand, the car will pump some of its stored energy back to the grid. Initially this will result in lower power bills, but over time it means that governments won't have to build new generation facilities as quickly as there is distributed generation capability—a very good argument for tax breaks on fitting solar cells to your home.

Lastly, technologies such as Organic and traditional LEDs, fluorescent light bulbs, ultracapacitors and other technologies are helping us improve energy efficiency today. The "incandescent" light bulb invented by Thomas Edison uses 4 per cent of the energy it draws to product light, the rest is converted to heat, or basically wasted. Organic LED TVs will draw a fraction of power that cathode ray tube (CRT) TVs used to draw, and even less than a fluorescent light bulb. **Low-power and energy efficient devices** will be an important trait in ethical consumerism of the next generation as individuals strive to be as close to carbon neutral as possible.

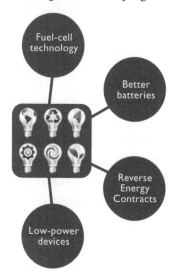

Enterprise-Wide Implications

So here is the kicker. By applying the **BANK** 2.0 paradigm, Moore's Law, Gilder's Law, Metcalfe's Law, along with the psychology of Maslow's hierarchy of needs, we get an unavoidable, unstoppable and unquestionable impact on adoption rates for innovation and new technologies. For the last 100 years or more, adoption rates have been reducing. The emergence of Moore's Law led these adoption cycles to pace up, then the Web again reduced the cycle. Mobile smart devices, always-on IP connectivity and wireless access have further compressed the adoption cycle.

While it is conceivable that adoption cycles cannot reduce down to nothing (i.e instant adoption), the fact is that not even the largest bank in the world could ever hope to slow adoption rates. For example, when Internet commerce started to emerge in 1996 and 97, many bankers and

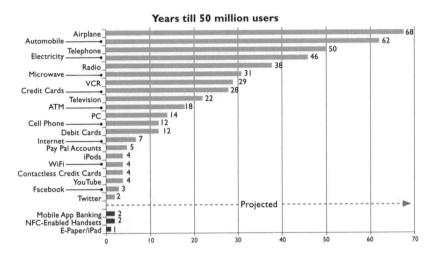

Figure 9.7 The shortening adoption cycles related to innovation

traditionalists in business either honestly felt that this was a fad, or they were so significantly threatened by the new "challenger" that they tried very hard to convince all and sundry that it was really a fad. And some argued it very passionately too.

> "Visionaries see a future of telecommuting workers, interactive libraries and multimedia classrooms … Commerce and business will shift from offices and malls to networks and modems … Baloney. Do our computer pundits lack all common sense? The truth is no online database will replace your daily newspaper, no CD-ROM can take the place of a competent teacher and no computer network will change the way government works … Yet Nicholas Negroponte, director of the MIT Media Lab, predicts that we'll soon buy books and newspapers straight over the Internet. Uh, sure."
>
> Clifford Stohl, *Newsweek*, 27 February, 1995[4]

Or this telling quote from a reformed banker:

> "I thought in rural Tennessee we would not be confronted with Internet banking in my lifetime. I was wrong."
>
> John L. Campbell, CEO of First Community Bank
> of East Tennessee, 1997

And, remember this one from an earlier chapter?

> "The do-it-yourself model of investing, centered on Internet trading, should be regarded as a serious threat to Americans' financial lives."
>
> John "Launny" Steffens,
> Vice-Chairman of Merrill Lynch, September 1998[5]

Right now in the long tail of perhaps the greatest economic crisis since the Great Depression, many banks are facing the full force of the bailout fallout fiasco. Banks around the world, which received billions of dollars in assistance to "keep the financial system going", are under severe pressure over bonuses and practices. The question though is, should we be so keen to keep the traditional banking system going? Are the bailout funds we've already expended on the financial system tantamount to trying to shovel out a snow drift while watching an avalanche descend from on high?

The fact is we cannot keep the traditional retail banking system going because it is facing something more destructive than the global financial crisis. It is facing something more significant in its progress for potential than the Internet "fad". **We are facing a complete re-engineering of the customer behavioural dynamic and a complete change to financial interactions at a consumer level.** A world where traditional banking with branches, ATMs, cash, cheques and the same old, same old, simply won't cut it. Why? Because the rate of change is speeding up, and no one and no organisation are immune.

Long held institutions and gentlemen such as Clifford Stohl of *Newsweek* and John Steffens of Merrill are probably still out there saying that the rate of change is slowing down. The detractors will be saying right now that we are just at the emergence of a new dawn where things will inevitably return to "normal".

It ain't ever going to return to "normal" folks. We are in the new **BANK 2.0** paradigm. Constant and unrelenting change based on the increasing capability and availability of new technologies and customers who just can't get enough of these new gimmicks. This is the new world order.

Banks which rely on branch network, cash distribution, cheque handling and other similarly outdated modes of interaction will simply become less and less relevant. Guess what … this is where we are at today, not in 10 years' time. For example, if you aren't already planning how you are going to phase customers off the use of cheques, you are going to be in trouble next year.

Now, the reality is banks are *not* going to disappear. However, their role in the day-to-day operations of the financial system will change.

If banks are no longer the mechanism for cash distribution and cheque processing, what is their role? Financial products, credit mechanisms, repositories of money, etc. all remain. However, the distribution system is completely disrupted by technology. As Mr Gates said, "Banking might be necessary, but banks … we're not really sure about those." Given that third-party players are increasingly infringing on the traditional banking infrastructure, banks may have to fight for their very existence. PayPal is a great example of this. Banks today still generally refuse even to acknowledge PayPal's existence, and yet it is the number one payments platform used on the Internet today.

So banks need to become very, very good at being virtual and digital repositories of their client's money, allowing access to that money anywhere at any time. Most of all, however, banks need to be great service organisations because the third-party challengers that are nipping at their heels will invariably be faster, more adaptable and more in tune with their customers.

Constant innovation, experimentation and testing of new ways and means to engage customers are the only things that will save banks from being replaced by more relevant mechanisms that utilise consumer will and device evolution in the coming years. If I sound like a broken record, please forgive me.

KEY LESSONS

Technology appears to have reached a point where exponential improvements are nothing new. Adoption rates for new technologies are skyrocketing and prices are coming down so more people have access to these technologies.

New materials science such as organic LEDs, nanotechnology and flexible circuit boards and chips are producing entirely new possibilities in device construction and design. As wireless technologies become increasingly pervasive, our devices are more intelligent because of their access to networks.

New interface mechanisms, augmented reality and increasing use of technology embedded into our everyday experience will continue to make new technology advances just the "norm" of doing business. Technology is not exceptional, nor is it an alternative choice for consumers—it is the way we do our banking in the BANK 2.0 world.

What will such a future bring? How does it impact service providers in the finance space? How can they prepare for it?

Keywords   Moore's Law, Adoption Rate, Energy, Fuel Cell, Battery, Flexible OLED, e-Paper, Nanotechnology, Quantum Computing, DNA Computing

Endnotes

1   Excerpts from *A Conversation with Gordon Moore: Moore's Law*, Intel Corporation, 205. pp. 1

2   SonyInsider.com

3   Wikipedia article "Gyricon"

4   "The Internet? Bah! – Hype alert: Why cyberspace isn't, and will never be, nirvana," Clifford Stoll, *Newsweek*, 27 February 1995

5   A widely reported quote from Merrill's John "Launny" Steffens during a presentation on the dangers of e-trading at a PC expo in San Francisco, September 1998

10 Gridless Customer Experience —More Complexity, More Choice

YOU'VE GOT MAIL! A catch phrase immortalised by the movie of the same name, and AOL's incessant audio tag associated with incoming emails. Email is one of those fundamental technologies that have changed our behaviour over the last 10 years or so.

To illustrate how our "experiences" have changed in the last decade in interaction, consider what percentage of our communication has been replaced by email, versus traditional mechanisms we customarily used such as the telephone, face to face meetings, letter writing and faxes. You'd be surprised. While email has replaced a lot of what we used to do, it has also added to the complexity of our day-to-day communications. Today we often just get more communication choices than we did back in say, 1990, rather than having emails replacing old methods outright.

Your Inbox Needs to Change for the Better

I want you to think about your Inbox content for a moment and how that has changed in the last five or six years. Now with Facebook, Twitter, Xing, LinkedIn, Plaxo, Naymz—all these various Web 2.0 technologies—we've increased even further the complexity of our daily interactions.

Take a look at my Inbox traffic mix over the last 30 days (Figure 10.1). By the way, this does not include junk mail, as this generally goes to my spam filter or I manually filter it out by deleting it. Spam and so-called malware would probably amount to about 80 per cent of my email otherwise. I have two prominent email addresses (two sets of company emails), and two secondary ones (Gmail and Hotmail) that I use regularly. These emails are split over four different clients or interfaces, but the summary shown

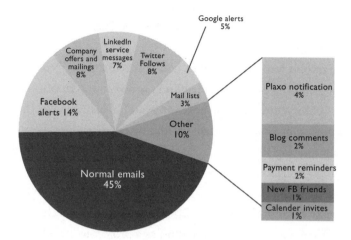

Figure 10.1 The author's inbox traffic for 30 days

includes only the priority email to my company email addresses which I access frequently each day.

Apart from email, I also have two other prominent communication mediums with business contacts—Skype and SMS. Ideally, I'd like to be able to keep IM (Instant Messaging) history for business conversations and emails from business contacts together, because often the email is a response to the IM conversation or vice-versa. But that isn't possible right now, although I sometimes find myself copying Skype conversations into an email to continue a conversation.

In addition, a lot of information I receive daily is in the form of social networking status updates, invoice reminders and bill payments, calendar meeting requests from business contacts and sales messages. Even with some fairly good filtering technologies, the fact is that I can't really organise my email effectively on the fly these days without spending a considerable amount of time manually sorting out this type of data. There are two ways that this is likely to be resolved.

The first is the creation of **supplemental dashboards** or overlays that act as a high level summary of information classified by sender type, relationship to you, or priority. Personal versus business contacts could be a further filtering trigger, although as social networks continue to take root, the lines between friend and business contact blur a little. These would need to integrate with both the IM clients you own and the nominated inboxes.

The second way that the inbox could be reformed is a completely new paradigm replacing the way we view and process conversations—essentially the 21st-century **unified conversation**(ware). This is not in the sense of unified messaging that is bandied around by telco vendors. We are talking about interpreting information that comes to you on a day-to-day basis and enabling collaborative real time responses.

One company that is doing something about this is Google. A brand new initiative from Google still in its beta stage is a technology called **Google Wave.** Google Wave started off as a project to see what email, if it was designed today, would look like, and it has produced something that is revolutionary. On its website, Google describes Google Wave as:

> "... an online tool for real-time communication and collaboration. A wave can be both a conversation and a document where people can discuss and work together using richly formatted text, photos, videos, maps, and more."[1]

Google Wave combines the best elements of a document such as a Word file or an email, with the interactions of a conversation. Email is after all essentially a conversation, but today we use things such as IM and SMS to supplement email because we need quicker response times. However, even in IM engagements, we really spend about 20–30 per cent of our time waiting for a response from our friend at the other end. So, even IM could be fairly easily improved in this sense.

A World Without Travel Agents and Stockbrokers, But With Virtual Agents

When was the last time you went to see a travel agent or a stockbroker? The age of the Web and direct channels has eliminated low value-add intermediaries. As customers become more empowered through the Internet and with the availability of detailed information in the pre-purchase phase, the role of the "advisor" has been relegated to one of absolute specialisation. Stock trading brokerage firms and travel agents are prime examples of this. While such intermediaries will not completely disappear, the only ones that survive the increasing margin squeeze will be those that can charge a premium for their role as a specialist advisor.

This increasing use of social networking tools will also mean that bank customers will become increasingly comfortable with virtual interactions. To illustrate this trend: since 2007, DoCoMo's iMode FOMA mobile phone now allows users to choose an avatar (virtual representation) that speaks for them in a video conference. Very useful when you've just stepped out of the shower and your video phone rings.

Avatars replacing IVRs

In most cases, avatars today are used as icons or profile pictures on Yahoo, MSN or Facebook. There are a bunch of sites and social networking tools that allow you to create some pretty amazing avatars of yourself from South Park's website, to the Burger King-sponsored Simpsonizeme, Zwinky, Yahoo! Avatar, MakeWee, Meez, Picasso, Mangatars, Toonlet, DoppelMe, IMVU, Second Life and many, many more. But the role of avatars is about to get a whole lot more serious.

As we discussed in early chapters, IVR (Interactive Voice Response) systems are increasingly being equipped with voice recognition technology so that natural speech can be accepted instead of touch tones from the phone keypad. The logical extension of this technology married with avatars are automated customer service representatives that will look and sound like a real person and be able to answer simple "canned" questions and respond to issues such as "What is my account balance?", "I've lost my credit card", "I need a new cheque book." Why might bank customers accept such an avatar simulated CSR experience? Primarily, that is because

Figure 10.2 The author's avatar alter-ego, South Park style
(Credit: South Park Studios Avatar Constructor)

it won't feel like you're talking to a computer; it will feel more like a human experience.

Says Mike Danielson, who is spearheading the avatar project at Motorola Labs: "As graphics hardware in phones becomes more powerful, more realistic avatars will be possible on cellular handsets. On a phone with the right hardware we can expect avatar display quality that exceeds what normally appears in a 3-D game, because we will only be viewing one character at a time."[2]

What IVR will become by 2015

Here's what your IVR experience might look like in the next five years based on the combination of avatars, voice recognition technology and predictive response modelling.

Good morning Mr. Green, how can I help you?
<Automated Video Avatar>

"I'd like my account balance please." <Mr. Green>

Which account is that for, Mr. Green? Savings, US dollar savings, or Euro savings?

"Savings"

The balance of your savings account is 320,422 Hong Kong Dollars. Can I help you with anything else?

"Yes I'd like to discuss a personal loan."

Figure 10.3 iPhone avatar
(Credit: iTunes, Apple.com)

You are pre-approved for a line of credit of 150,000 Hong Kong Dollars at an interest rate of 5.25%. Would you like to accept this offer?

"No, thank you."

Let me transfer you to a loans advisor so you can discuss your requirements. Thank you for calling your bank.

Touchy Feely

The biggest single improvement in screens and devices over the last few years, however, is obviously the integration of multitouch, touch-screen capabilities. Of course, we've had touch-screen technology for many years, but not touch screens that allow you to use multitouch or multiple fingers on the screen simultaneously. Multitouch enables unique actions such as pinch, poke, swipe and pan that were not possible with previous iterations of touch-screen which could only cope with one digit interaction at a time. This is a fairly recent technology and comes through the iPhone, Microsoft Surface and other similar devices.

More recently, we've been hearing a lot about "haptic touch" emerging so that touch screens and devices will provide more interaction capability. Haptic feedback, for example, would allow you to use an on-screen keyboard that feels like a real keyboard in its response to your touch. It accomplishes this by using vibration technologies similar to the motors that are activated when your phone is on vibrate mode. Apple is reportedly releasing a haptic feedback, multitouch "mighty mouse" as a replacement for their current Mac mouse series.[3]

The one failure we discussed on the iPhone design back in Chapter 6 is the poor comparative usability of the on-screen keyboard, which has an unusually high error rate compared with its RIM competitor. Haptics may work as a mechanism to resolve those issues effectively. If you feel like you are using a real keyboard as a result of haptic feedback, then the keyboard (and the user) will behave as if it is a real keyboard, and accuracy will be improved dramatically.

This is why it is likely that the mouse and physical keyboard will disappear over the next 10 years. Screens will become the PC or laptop, your fingers will be the mouse, and an on-screen haptic keyboard will be there when you need to type.

As a result, screens will get larger, and for "desktop" PCs, we will more than likely use something more akin to drafting tables that sit on an angle than the current screens and CPUs we use. As we do more things wirelessly via LAN and as hard disks become solid state, we will do away with DVD drives for the most part, so our PCs will be contained completely within our screen effectively.

The iPhone started this all, but it is by no means the finish. The next seven to ten years will really produce a very significant revolution in these technologies. So we have to be ready to use this and produce applications that leverage these capabilities.

Computing in the Cloud

As we become more mobile, a great deal more of what we do will need to become detached from our work computer, laptop or enterprise network server. The ability to get access to our data and core applications on the move is one simple example. Restricting this data to physical devices in one specific location is not going to work. So the trend has been for laptops to get more capable so that we can carry them with us. But laptops still have to deal with access to corporate data, security issues and such, regardless of their portability.

For this reason, Google, IBM, Apple, and even Microsoft are making various bets on what is known as **cloud computing.** Cloud computing is an emerging computing technology that uses the Internet and central remote servers to maintain data and applications. It allows the use of applications without installation, and allows users to get their personal data and files using any device that has Internet access. Cloud computing abstracts users from their applications and data by providing those facilities via the browser effectively, making storage requirements minimal and leaving processing to the cloud rather than requiring heavy processing capability. It does, however, rely very heavily on bandwidth to get expeditious results.

Although Windows still runs on 90 per cent of PCs and laptops, the fading importance of the PC means that Microsoft is no longer all-powerful. While some look to Google's Chrome OS and Apple's OS to produce a viable competitor to Windows, the fact is that with increasing reliance on mobile devices, app phones and other device platforms, the role of the PC operating system is not necessarily a key driver in the future of computing.

While there are hundreds of firms offering cloud services (Web-based applications living in data centres), Microsoft, Google, IBM and Apple play in a league of their own. Each of these firms has its own global network of data centres and they are working on a whole suite of services within the "cloud". Such tools and services being touted include email, address books, storage, collaboration tools and business applications. App phones or smart devices will add to the mix, with various approaches to widgets and applications, some from within the browser and others integrating with the cloud.

Apple has recently started to make a foray into cloud computing in a major way. Apple is building a $1 billion data centre in North Carolina, US, possibly the largest of any in the world. MobileMe® is the first of a series of online services based on cloud computing that is designed to create new revenue streams for the tech giant. MobileMe is designed to connect all your devices and push information up and down to keep everything synced and up to date. iDisk, incorporated into MobileMe, gives you 20GB of remote hard disk space for storing files that are too big to email, photo galleries, and such. MobileMe even allows you to find your iPhone if you've lost it.

While Microsoft has launched Office Live as an attempt to win over cloud enthusiasts, their poor mobile showing in recent times against both Apple and Google's Android might affect their ability to dominate the cloud as they have the PC market. Motorola's ditching of Microsoft's Mobile in favour of Google's Android platform may just be another nail in the MS Mobile coffin.

The question remains as to what services work, and what revenue models will drive cloud computing. For corporations, the business case is simple; shifting to the cloud reduces infrastructure costs and moves platform and application costs to an OpEx (Operating Expense) model instead of CapEx (Capital Expense). In the current economic environment this has to be promising. Distributed platform access and the benefit of data centres in the cloud also create more opportunities for more agile institution operations and different models such as telecommuting, homeshoring, pop-up branches and so forth.

If you are sitting there reading this right now with some scepticism about the possibilities of the cloud, think about this. Arguably the most successful cloud computing service today, with more than 350 million users, is **Facebook.** If you can't get your head around that, think about it. It is run completely through your browser or app widgets. It allows users to collaborate, share information and communicate online—all things that businesses want to do to.

Whether Google, Apple, IBM, Microsoft or another contender such as Amazon or Facebook is able to dominate the cloud space is not really the issue. The issue is that if you are a decision maker in a business, you need to think about whether some of your core infrastructure, platform or applications would be better placed in the cloud so that your workforce can be more innovative and productive. It can be just as secure as your own dedicated infrastructure, plus you get the benefit of much more mature user interface and shared services.

One of the more interesting elements to think about with cloud computing is what happens to the role of the bank with regard to payments. If the majority of payments processes on a consumer and B2B space are increasingly deployed within the cloud, then the need for such services provided by banks to their customers directly becomes effectively redundant. After all, PayPal is already in the cloud.

Widgets

Consumers are becoming comfortable with transactional widgets, meaning banks must get more comfortable with Web 2.0. A software widget is a generic type of software application and is usually portable and frequently written OS-agnostically. The term "widget" often implies that either the application, user interface, or both, are light, meaning relatively simple and easy to use. Examples of widgets may be an OS desktop accessory or applet, as opposed to a more complete software package such as a spreadsheet or word processor. In recent times, widgets have become a major class of applications in use by the broad consumer base. Since the launch of Vista, Microsoft has supported "gadgets" running either through Sidebar or independently in Windows 7.

A survey by Web 2.0 start-up Worklight found that 63 per cent of the 500 financial professionals surveyed identify transaction widgets as a priority for 2010.[4] Transaction widgets might allow customers to check their balance, transfer funds and pay bills directly from their desktops, mobile applications, or social networking sites such as Facebook.

Increasingly we are going to see iPhone apps, Android apps, Microsoft Gadgets, Ovi apps, Apple Widgets and other such application platforms evolve into small, portable pieces of software. Widgets will be an important move to cloud computing too, as more and more users seek to build interfaces to their life.

Yodlee is the engine behind the online banking operations of many of the largest banks in America. In recent years, Yodlee has built up an enormous dataset—more than US$3 trillion of transactions from 23 million users have been cleaned up and put into a huge database.[5] Yodlee has recently opened up that database to software developers around the world for the construction of applications, or "widgets". These widgets will allow people to do things with their banking relationship or in managing their personal finances, which until now simply hasn't been possible.

For instance, a billing company might build a widget to look at all your bank accounts and give impartial advice on where you might be losing money, or match you to a utility vendor that is a better fit for your usage patterns. New payments widgets could appear on the scene, using the cheap Automated Clearing House (ACH) network rather than expensive wire transfers, allowing people to pay each other easily without going via PayPal. Your financial advisor or aggregators could set up a simple dynamic comparison tool showing how the various bank rates and fees compare. Developers such as Josh Reich of i2pi are talking about building apps which let you take a photo of a bill with your iPhone, and then allowing the widget to pay the bill automatically on the due date by coordinating with your bank.

PayPal filled a gap in payments infrastructure because Visa, MasterCard, and the banks all moved too slowly to capture this space. MoBank has filled a gap in balance enquiry capability because most banks have moved too slowly to enable this on the iPhone platform. Just like PayPal and MoBank, the abundance of cheap and very effective widgets may chip away at thousands of revenue opportunities for the banks of today.

How many widgets or applications like this are banks building today? If they are building one or two, then they are particularly advanced, but it is still nowhere near enough to fend off the gaggle of third-party developers entering the fray. This picture is wrong.

Banks need to be engaging communities of developers to assist them in providing more functionality to customers. However, banks will typically keep third parties completely away from access to their back-end systems out of fear of security and privacy issues. Banks of the **BANK** 2.0 paradigm need to take a leaf out of Wikipedia in respect to article development and authoring for widgets. Providing developers with sandboxes to develop widgets in return for either small subscription fees or some sort of fee-for-performance in customer take-up and/or use could work. Why not? It worked for Apple. What have the banks got to lose? Just their customers.

Interacting With Your Environment

Something that is a little bit out there but interesting to think about, is the emerging technology around image recognition. We've had OCR (Optical Character Recognition) for many years now, but there have been recent improvements in image processing and matching. Recently, Google has developed search engine technology called Google Goggles that allows users to search based on images taken by your camera phone. It currently is in beta with some reasonable search support for books, DVDs, landmarks, logos, contact information, artwork, businesses, products, barcodes and text. It won't be long before this is perfected.

Combining this type of technology with digital cameras or camera phones is one thing, but there is an emerging technology that might change the way we see our environment and the things around us in an entirely new manner.

Smart glasses and contact lenses

At Sony's 2009 Consumer Electronics Show, Tom Hanks appeared on stage with Sir Howard Stringer, CEO and president of Sony Corporation in the US. Sony was parading its new high definition video glasses that are currently under development. These HD specs have a widescreen 16:9 HD Quality image projected onto the lens. In the show version, they also had in-built camera(s).

In 2003, MIT published a paper on the concept of smart glasses they called "The Memory Glasses".[6] These memory glasses used both cameras and FLIR (Forward Looking Infra-Red) for image recognition and a (HUD) Head-Up-Display system for visual cues. The prototype glasses were interfaced to a large database of objects that could be recognised by the glasses. Facial recognition software could be used in tandem with the glasses to help you recall an acquaintance's name or specific details. MIT experimented with both overt and subliminal cues for proactive memory support using the glasses, showing the application for those suffering memory loss, Alzheimer's, or amnesia.

Recent advances in nanotechnology applications have enabled the creation of prototype contact lenses with a built-in pressure sensor using a novel process that etches tiny electrical circuits within a soft polymer material. In other research, scientists at Boston Retinal Implant Project have been developing a bionic eye implant that could restore the eyesight of people who suffer from age-related blindness. The combination of these various technologies leads to the conceptualisation of some very interesting innovations.

Augmented reality

Augmented reality (AR) is the term for real-time digitally enhanced interactions with the physical real-world environment, where real-world elements are merged with (or augmented by) virtual computer-generated imagery, touch or positive feedback, sounds and, even possibly, smells. The resultant mixed reality is what we term "augmented". The term augmented reality is believed to have been coined first by Thomas Caudell, an employee of Boeing, in the early 1990s.[7]

Augmented reality is changing the way we view the world—or at least the way tech users see the world. Picture yourself walking or driving down the street. With augmented reality smart displays, which will eventually look much like a normal pair of glasses, informative graphics will appear in your field of view and audio cues will provide information or feedback in respect of whatever you see.

Applications of smart glasses could be anything from an equivalent of your current laptop display while you are on the move, to simply a Bluetooth plug in your app phone showing you in real time a virtual Head-Up-Display with key information from your device such as Caller Id, local weather, e-alerts or appointments. Incorporating image and facial recognition software, along with RFID technology, smart glasses could remind you of the name and details of a key business contact, an old school friend who passes you by while you're chilling at the mall, or the current price on Amazon of that book you're looking at through a retailer's window. The possibilities are far-reaching, and just a little freaky.

In any case, within just five years we could have access to such devices married to our app phones or while watching a movie or receiving a video phone call. That is all pretty amazing. Glasses could become the next iPhone-type fashion accessory. Right now, both iPhone and Google Nexus phones incorporate some AR applications that are very simple to use and very, very cool.

Figure 10.4 Where's my nearest NY subway station? (Credit: Apple.com)

KEY LESSONS

As already borne out by our review of technology adoption, along with Web 2.0, the app phone phenomenon and so forth, customer experience is now in a state of gradual, but constant change.

What is going to happen next to our customer experience? The way we receive, prioritise and review information will have to change. We are simply getting too much of this information now to deal with it the way we always have.

Secondly, the way we interact with technology is going through an evolution. Touch screens are allowing our fingers to replace the mouse and keyboard, and new ways of accessing data, user interface approaches and application platforms are opening whole new ways of organising, processing and prioritising key content. Augmented reality is changing the way our devices will interact with our environment.

One thing is certain, however. Customers and the financial institutions that serve them will be adapting continuously over the next decade

Keywords Customer Experience, Haptic Touch, Augmented Reality, Virtual Worlds, Avatar, Google Wave, Cloud Computing, Widgets, Google Goggles, Smart Glasses

Endnotes

1   Google.com

2   "Building Your Own Phone Face", by Reena Jana, *BusinessWeek*, 19 January 2006

3   Geek.com, "Apple has a mightier mouse that needn't be moved at all", 5 October 2009

4   Worklight Survey of Financial Professionals, 4 November 2009

5   Yodlee.com

6   "The Memory Glasses", MIT, Richard W. DeVaul – ISWC 2003

7   Wikipedia.com article: "Augmented Reality"

11 The Emergence of the Prosumer—Collective Intelligence, Social Networking and Web 2.0

Are You Ready For Your Second Life?

HAVE you heard of Linden dollars? Linden dollars are the official currency of Second Life, an online virtual world with millions of members and an economy that's growing by 20 per cent a month. Named after Linden Lab, the company that developed Second Life, the Linden dollar is completely exchangeable with the US dollar at a rate of around L$280 to US$1. Currently, over $1.5 million in transactions (in US$) occur on Second Life every day as members buy virtual cars, real estate, gadgets, and just about anything else you can find in the real world, as well as a lot of things you can't—your very own ninja avatar, for instance.

The idea of Second Life is just that—you can have your second life in a virtual world, just the way you like it. In Second Life, you can be a rock star, you can own a space station, you can completely tailor the way you look. It all takes place online in a virtual world. ABN Amro and Fortis Bank think this virtual world is real enough to offer an advisory service in the virtual space. Yes, you can walk into this branch in Second Life and ask for financial planning advice.

Second Life is one of many online social networks taking the Web by storm currently. Others include Facebook, Twitter, LinkedIn, YouTube, Flickr, Digg, and many, many more (over 2,000 by the latest count). These allow users to network and interact in ways that only the Internet makes possible. Sharing photos instantly with my family on the other side of the world, finding a business associate who recommends a great media firm in Shanghai, publishing my own newspaper column on my blog, or looking for love—it's all available online with the Web 2.0 phenomenon.

Figure 11.1 The author in his Second Life avatar-guise visiting Bradesco Bank in virtual space (Credit: Linden Lab)

The Prosumer— Why Web 2.0 Works

Many of the top 10 jobs in 2009 are jobs that didn't even exist in 2000. Here are some examples:

- Tech jobs (iPhone application developer, widget developer, social networking architect, data communication analyst).
- Environmental engineer (reduce carbon emissions).
- Cosmetologist or epidemiologist (botox and dermal filling).
- Digital advertising executive (new media).
- Forensic accountant.

We are preparing students now for jobs that don't yet exist. Much of what they are taught will probably be redundant or out of date by the time they graduate. The US Department of Labor estimates that today's learner will have 10–14 different jobs by the age of 38.[1] Web 2.0 is changing the dynamic of how we live. It's not just with our work either. One couple out of every eight who married last year in the US met online.

Each day, we receive more and more information to process as a result of the information age. The number of text messages sent daily exceeds the total population of the planet. Indeed, during the 2009 Lunar New Year in China, more than five billion text messages were sent in the space of just one week. It is estimated that a week's worth of the *New York Times* newspaper contains more information than a person was likely to come across in their entire lifetime in the 18th century. As a result of all of this content, some 5 Exabytes (that's 50,000,000,000,000,000,000 bytes) of unique information was generated in 2002 alone.[2] This was more than was collectively generated in the preceding 5,000 years. To top it off, the amount of technical information generated globally is now doubling every two years. Where is all this content coming from?

Introducing the **prosumer,** a producer and consumer of information. The prosumer is you—a Web user that creates and uses available content and information on the Web. The Web 2.0 phenomenon is responsible for the creation of user collaboration that makes social networking, virtual worlds, networked applications and other such tools work. Today, we are no longer just a consumer or "reader" of content. Today, we are producing and consuming content at a voracious rate. Here are a few statistics of the prosumer in action:

- Every minute, 20 hours of video are uploaded to YouTube.
- Every minute, 14,564 tweets are made on Twitter.
- Every hour, approximately 10.5 million songs songs are illegally downloaded mostly from areas where legal downloads of the same are not possible (250 million songs a day).
- Every day, 2,300 new Wikipedia articles are created.
- Every day, more than 175 million Facebook users log on to FB.

We are facing an informational revolution. With the pure amount of information we have to deal with, we need to find new tools and mediums for organising this information. Ironically, Web 2.0 has created both the tools for generating this content, as well as managing it in many instances, for example, YouTube. Given that so much of the information we generate is dealt with in cyberspace, a great deal of what we receive in the real world is becoming bothersome because as content it is more difficult to deal with. Hence, traditional mail is now derided as "snail mail". Traditional print media is failing, and even TV is being displaced by the Web as the leading provider of content.

This shift means that banks and other organisations need to be delivering content through this new medium or they become irrelevant. But right now, today, surveys show that more than 50 per cent of banks are so threatened by this new phenomenon that they have banned the use of social networks at work.[3] Customers are very willing to collaborate with their bank and other service organisations to improve the whole experience, but by not participating in these forums and networks, banks are missing out on these opportunities. This is simply further proof of how out of touch with the **BANK** 2.0 paradigm most banks are.

Let's look at two reasons why the prosumer has such a profound effect on our lives in the last few years. The first is a "law" that defines why networks grow as they do. The second is a look at the phenomenon of social networks.

Metcalfe's Law—the nature of networks

Metcalfe's Law is attributed to the founder of 3Com, Bob Metcalfe. Metcalfe's Law characterises the network effects of communication technologies and networks such as the Internet, social networks, and the World Wide Web itself. Metcalfe's Law is not an iron clad empirical rule like Moore's Law, but as a heuristic or metaphor, it is extremely powerful in describing the reason some networks grow so quickly and exert so much influence. Moore's Law has been numerically accurate since 1967. Metcalfe's Law, on the other hand, has never been evaluated numerically. It states:

> "The value of a telecommunications network is proportional to the square of the number of connected users of the system (n2)."
> Robert Metcalfe, founder of 3COM, 1980[4]

Sounds too mathematical for you? Let's put it in very simple terms. If you link two computers or two telephones together, they can make only one connection. Five computers, however, can make 10 connections, and 12 computers can make 66 connections. The law has often been illustrated using the example of fax machines. A single fax machine is useless, but the value of every fax machine increases with the total number of fax machines in the network, because the total number of people with whom each user may send and receive documents increases as the network grows.

Metcalfe's Law calculates the "value" of a network, but as such the mathematical formula is a justification or measure of the potential number of contacts. For example, even though I can send a fax to you, it doesn't mean I actually will.

The **utility** of a network, however, depends on the number of nodes in the network that are actually in contact with each other. This defines why QQ messenger is so phenomenally successful in China, but is virtually unheard of outside of China. If non-Chinese users can't communicate with the users of QQ messenger because of language issues, the utility of the network of those who don't speak Putonghua/Mandarin is virtually zero.

There are other factors affecting utility also. Chris Anderson popularised the concept of the **long tail** in an October 2004 *Wired* magazine article, showing the value of networks for hard-to-find items. This further describes value creation of broad networks at the low-value end of the spectrum, such as in the application of Muhammad Yunus' Grameen bank, and Amazon or Netflix.

What makes social networks so successful?

In Malcom Gladwell's book *The Tipping Point*, he refers to super connected people (who are enablers for all sorts of business, interactions, and so forth) as **connectors.** What we are seeing with Web 2.0 tools such as Facebook, LinkedIn and Twitter is that these are acting as proxies for connectors and are enabling networking connections in many new ways.

We are social networking every day. If you tell a friend about a service or product you like—you're social networking. If you hand out a business card—you're social networking. When you return phone calls, respond to email, blog, write on forums—all this falls under social networking.

Social network groups are not new either. There are already global networks such as Toastmasters, Business Network International, Million Dollar Roundtable and Globond. The American Chamber of Commerce runs a great offshore networking system, and even Multilevel Networking Systems (MLNs) such as Amway create global networks to further growth. These networks all need an organisation, infrastructure or system to make them ultimately successful. As soon as someone stops organising the events or sending out the newsletters, these types of networks start to wane.

People like to be around other people with similar ideals and goals. Business social networks work because they foster common goals, or have participants with common interests and focus. If you are going to get involved in one of these networks, however, work out very specifically what you want that network to accomplish for you. Create a set of goals. Don't get trapped into only helping others with their goals, otherwise you will lose your focus. That said, you do need to contribute to the network to get something back for yourself. It's a sort of group karma thing.

Online social networks work to mimic the best parts of real world social networks, but have a few defining elements that make them more successful. Firstly, online networks are easy to join; there is no hurdle, no joining fee (generally) and no restrictions on entry. Secondly, growth is not limited by physical space or geographic location; thus, these networks tend to grow very rapidly once they get into the psyche of the intended user. Lastly, like real world networks, you get out of them what you put into them, but if you don't want to do anything you can still stick around.

I've heard many senior executives with banks justify to me why they aren't on LinkedIn, Facebook, Twitter and other social networks. The primary justification I hear is that they get bothered by too many requests. If you're an executive in retail financial services, you might feel the same. The last thing you need is another vendor tracking you down, wanting to know if you could get them an introduction to your head of IT, or something similar.

For a variety of reasons, social networking has been slower to take off in the business world. Employees are wary of disclosing too much to potential competitors, and loose-lipped executives can easily embarrass themselves and their companies online. Also, business users typically have less time to devote to socialising online and are willing to do so only if they believe they are getting a unique benefit from the site.

This is one point of view. Other senior executives take the view that they can learn a great deal from participation in these networks. First of all, it gives you direct access to your customers in a way that is unfiltered and unbiased by the "spin" of marketing research conducted by your own marketing department. Secondly, by understanding the way these social networks function, you can understand the ability of these networks to create new markets for your organisation.

There are times when social networks can be very powerful for an organisation. You might recall that during the recent Iran elections, Twitter was used to great effect to identify the irregularities in the process. As a PR tool, it worked far better than anything else the incumbent could produce online. In February 2008, in the run-up to the US elections, John McCain raised US$11 million through campaign fundraisers to support his presidential bid. Barack Obama attended no campaign fundraisers, but used online social networks to raise $55 million in just 29 days.

Other specific social networks are popping up for dedicated networks of professionals in specific fields. These include Within3.com or Sermo. com (for medical professionals), InMobile.org (for executives in the wireless field), DiamondLounge.com (a sort of excusive network dating service for successful executives and billed as the social network that doesn't want you), and ReutersSpace (for professionals in the trading field).

Social networking is not just a tool for geeks though. Its use for executives is an important litmus test for innovation within an organisation. Professor William Baker of San Diego State University surveyed 1,600 executives and found that firms that use external social networks scored 24 per cent higher on a measure of radical innovation than companies that don't.[5] To illustrate this fact, Dell has reported on its corporate blog that it has already made more than US$3 million from sales through the Twitter network.[6]

Living with increased transparency

As the amazing Twitter response to Michael Jackson's death and the Iranian elections show, the old saying "Good news travels fast, bad news travels faster" has never been more true. A recent book *Groundswell*, produced by two leading analysts from Forrester Research, puts it this way:

> "Right now, your customers are writing about your products on blogs and re-cutting your commercials on YouTube. They're defining you on Wikipedia and ganging up on you in social networking sites

like Facebook. These are all elements of a social phenomenon—the groundswell—that has created a permanent, long-lasting shift in the way the world works. Most companies see it as a threat."
Groundswell, Harvard Business School Press, April 2008

The social media space has literally transformed itself, with the average profile age on LinkedIn at 40 years, while the same for Twitter is close to 35 years. Facebook's biggest age group of followers is now in the 35–54 range, as per data from the site's management. Don't fall into the trap of thinking that those on these sites are not in your demographic. They ARE your demographic.

In the midst of the emerging global financial crisis and General Motor's demise, Ford did something completely unexpected. Scott Monty, the head of social media for Ford, approached Ford's CEO Alan Mulally and asked if he would answer some questions on Twitter. The response? "What's Twitter?" When Monty explained the concept to Mulally, he was apparently all for it, and in an interview conducted by Twitter via Monty's BlackBerry the results unfolded. It was the first time a Fortune 500 CEO had directly interacted via Twitter with an open audience.

Jeff Jarvis, author of *What would Google do?*, creator of BuzzMachine. com and famed American journalist who works fiercely in the blogosphere, gained national notoriety when he wrote about his negative experiences in dealing with Dell Computer's customer support system on the website. This and other pressures led Dell to take a much more proactive stance online through Twitter and other social networking tools, which we've already seen were very successful. Dell's strategy was simply to go out and fix any problem that was blogged about as quickly as possible. Of course, the bloggers all thought this was wonderful and started to sing Dell's praises in the blogosphere.

Others have not been so successful because they have missed the point. According to Jarvis, companies such as Sprint in the US have social media people out there responding to negative blogs by reinforcing policy and defending the company instead of proactively trying to resolve customer issues. He cites the example of problems he was having with his new Kindle e-book when he commented on his blog that Sprint network connectivity was the cause of his Kindle problems. Here is how Jarvis categorised the response by the Sprint CSR on Twitter:

> "When I went up the street [the Kindle] worked. Now Sprint has someone watching Twitter, but his response was so "Phone Company!" It was … 'Well it's not our problem; we didn't have a network outage. You should be calling Amazon … you were better off not Twittering fella!"
>
> Jeff Jarvis, Columbia Business School, 4 March 2009

Twitter, Facebook and such tools are valuable sources of customer information. These offer you insights into the problems customers are having. Now Jarvis argues, rightly, that Sprint would have been better focusing on investing and attempting to fix the problem, rather than justifying their mistakes. In the meantime, all of this is being played out online with lots of people watching Jarvis' YouTube clips and reading his blog on how Sprint responds to such issues.

Banks say they are establishing presences on social networking sites to tap into a growing demographic and to control the conversation about their brands. The economic crisis is undoubtedly another driver.

It's no coincidence that the hue and cry for corporate social responsibility and ethical consumerism is growing at the same time that corporate communication is being steadily democratised. We are moving from the patriarchal corporate identity ("what is good for GM is good for the US"), to flatter communication via the Internet (if still mostly advertorials), to increased transparency via employee and executive blogs, and finally, now a real bi-directional dialogue via social media.

> **"There's a lot of worry out there…That means that we have to stay close to our customers."**
>
> *Ed Terpening, Vice-President of Social Media, Wells Fargo*

How Will Web 2.0 and Social Networks Transcend and Transform Banking?

In a mature digital space, consumers will opt for personal loans or credit facilities from trusted online aggregators or brokers. Zopa.com is one of these emerging social networks.

Zopa is an online financial network that allows investors to lend to consumers/borrowers through an intermediary platform online. It offers borrowers the ability to get lower rates than through the banks, and investors to get higher deposit rates than they get through banks. Harvard is already touting Zopa as the next revolution in banking. But the management team has no banking experience, and the banks cry out "Doesn't Zopa need a banking licence?" Well, they don't actually take deposits; they just facilitate two people agreeing to enter a contractual arrangement that looks like a traditional personal loan, and for this they take a small fee as the intermediary.

Figure 11.2 Zopa.com, the non-banker lender making waves and profits (Credit: Zopa.co.uk)

In the US, Zopa decided to abandon the market because of increasing pressure from banks, the SEC (Securities Exchange Commision) and Congress, and instead stick with their European presence which was already paying dividends. Peer-to-peer lending companies, Prosper and Lending Club, are both heavily invested in the US market, however, so they chose to fight for the right for P2P lending networks. The lenders received welcome news from the House of Representatives in December 2009, when an amendment was added to the bill that would move the regulation of P2P lending companies from the SEC to a new regulatory agency known as the Consumer Financial Protection Agency (CFPA).

Previously, under pressure from banks and lobby groups, the SEC had claimed regulatory oversight of P2P lending companies such as Lending Club and Prosper, even to the point of forcing Prosper to shut down while going through a process to register its loans as securities. The regulatory debacle effectively brought the P2P market in the US to a standstill. Lending Club and Prosper both made it through the regulatory hurdles, but Zopa decided to pull out of the US market.

Patently, peer-to-peer lending is not a clean fit with SEC regulated securities. Under pressure from banks which did not like Web 2.0 start-ups messing with their markets, the SEC was forced to attempt to step into the breach and try to regulate these new lenders. Could it be that the issue was not that P2P lending was illegal, but that it had slipped in from under the radar of the big banks and they felt under significant threat?

If there was such a significant opportunity in this space, why didn't the banks think of this approach themselves? They probably did, but why build something that destroys margin on product, and requires eliminating silos and viewing risk in a whole new way—very difficult propositions for the average retail bank indeed. But it works for customers because it is more efficient and more cost effective than traditional banking. This is the lesson in Web 2.0. Social networks will find a way of getting around inefficiencies in the banking network, and they will look like very attractive propositions to customers. So we need to move faster to eliminate issues such as lengthy approval times, channel inefficiencies, marketing mix and contact management.

The customers of 2020 will have grown up with these technologies and will be used to dealing with the dynamic of social groups spread across the virtual space. These networks will be much more efficient at providing trusted advisory services than the bank, and they will compete with our traditional banking services head to head. Tim O'Reilly, founder of O'Reilley

Media, puts it well: "The essence of Web2.0 is that networked applications get better through participation. It is true collective intelligence."[7] He should know—he was the individual who coined the phrase **Web 2.0.**

Some banks have already started to embrace social networking. Citigroup has a business group on Facebook with over 5,500 members, and also has thousands of employees on the LinkedIn platform. Leading professional services firm Ernst and Young has over 80 separate groups on Facebook dedicated to employees, and of the other so-called Big Four firms, all have alumni groups and other such groups on Facebook too.

So how can an enterprise use social networking effectively? There are six key benefits of social networking that are frequently utilised by professionals in enterprise today.

(1) Collaboration

For research and for connecting with other like-minded professionals, social networks enable you to initiate contact with others in your field that you may otherwise never have met. Medical researchers, market researchers and others have told of being frequently surprised by individuals they've found in the cloud working on exactly the same research or with the same thinking.

(2) Mobilisation and supply chain

Sites such as Alibaba.com and others enable you to connect with suppliers and partners that have never before been possible, particularly as you reach into new global markets or require special, localised knowledge.

(3) Finding talent

Within three years, classified ads in newspapers for recruitment will virtually be gone due to the success of online recruitment, and by the time a job opening is posted, there's likely to be hundreds of resumes already in the queue. Apart from dedicated job sites such as Monster.com, the promotion of job opportunities through social networking sites such as LinkedIn, Facebook and Twitter is on fire. Why advertise through a newspaper when you can utilise social networks of people you know to find the right person for the job? Companies such as Goldman Sachs also offer videos of what it is like to work with the company and put these out on YouTube; others use podcasts too.

(4) Viral marketing

The most obvious benefit for enterprise is marketing. Social networking sites and networks can create great opportunities for marketing new products or even brand, as Dell has shown. But this requires dedicated resources that can create the messages and the buzz that extends the message into the broader domain.

(5) Community building

Banks are not usually known for building warm and fuzzy communities around their products and services. When we think of banks, we often think of impersonal bankers in pinstriped suits denying customers their request for a car loan or a mortgage. The world is changing, though, and even banks are trying to foster community rather than appear monolithic and imposing.

(6) Product research

Whether you're crowd-sourcing to find out what customers think of your services, or using social media as one tool in your arsenal to enlist customers to help develop new products, a social network is an undeniably powerful research and development resource. Enterprising banks are even using social media tools to look at their customers' needs and refer that into the process that creates new products and services.

New ways of using networking

Social networking is mobilising in new ways that we haven't even contemplated as yet. Part of that is the effect of the use of mobile devices to interact with networking tools.

TribeFinder™ is one of those new tools. TribeFinder is a technology that analyses "Peer-to-Peer" data from network operator data (Customer Data Records). From the CDRs, it identifies the *Naturally Formed Social*

Network Clusters—they call them **Tribes**—within the client base. These can be analysed by:

- number and frequency of communications per period;
- communication media such as Voice, SMS, MMS, IM;
- time of day communication takes place;
- location of caller/receiver;
- on Net versus off Net composition; and
- identifying the **"Key or Peer Influencers"** (the most interconnected individuals by frequency and reach) within each Naturally Formed Tribe.

Simply by analysing the calls and SMSes you make every day to your friends and colleagues, a sort of natural network of friends is formed—a tribe if you like. Key influencers are the power-networkers who can move or influence networks because their opinion counts. Individuals in the network follow these key influencers because they are trendsetters, sort of like Oprah with her book show. It won't be long before networking merges with customer analytics to enable marketers to target key influencers with viral messages that can flow out to these tribes or networks.

You send one SMS or MMS to a key influencer, and then they send it out to their tribe if it is perceived to be of group value—an exciting new social marketing approach that is yet to be tapped.

Endnotes

1   US Department of Labor. http://www.bls.gov/news.release/pdf/nlsov.pdf

2   "How Much Information?", UC Berkeley Study 2003. http://www2.sims.berkeley.edu/research/projects/how-much-info-2003

3   Sophos Security Online Survey

4   Guest blogger Bob Metcalfe: Metcalfe's Law Recurses Down the Long Tail of Social Networks

5   "Four Ways Social Networking Can Build Business", Jake Swearingen. BNet.com

6   Direct2Dell blog

7   Tim O'Reilly's blog – radar.oreilly.com/2005/10/web-20-compact-definition.html

12 Future Payments and Cash— RFID, Biometrics, P2P Micropayments, Digital Cash

Not Just Chip'd Cards, but Chip'd Everything

Conrad Chase wanted something unique to identify the VIP patrons of his Baja Beach Club. After brainstorming with friends and business associates, Chase came up with the idea to implant his VIP members with VeriChip's implantable microchip as a virtual credit card for utilising the club's facilities. The cost? €1,000 to implant the chip and get VIP status, plus €1,500 credit for your bar tab at the Baja. It was a huge success.

The RFID chip that is implanted is 8mm long and 1mm wide, and is enclosed in a glass vial. It can be implanted in minutes and is painless. A magnetic RFID reader can be simply passed over your shoulder where the chip is inserted to check your bar tab and to verify your membership in the exclusive club.

Some people believe that by 2020, parents may even choose to chip their kids at birth because the RFID chip of that generation will be capable of monitoring vital signs and health, and anticipating nasty bugs or health problems.

But if you are nervous about implanting a credit/ID card-type chip in your shoulder or wrist, do not despair. Why not simply wear it on your wrist embedded in your watch, or have the chip integrated in your

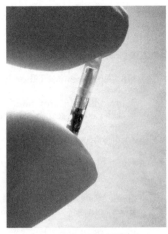

Figure 12.1 VeriChip's implantable microchip
(Credit: VeriChip)

clothes? The size and cost of RFID chip technology will mean that you can have your clothes or jewellery "encoded" to act as an interface with your virtual mobile wallet, so you don't have to carry any money on your person.

Biometrics and no more passwords

We've been hearing it for years and watching it on sci-fi movies such as *Minority Report*, but human-computer interaction (HCI) technologies and biometrics are becoming more ubiquitous every day.

For example, fingerprint-enabled passports or ID cards are now used in the EU, Hong Kong, Dubai and elsewhere every day. We can also use fingerprints to access our laptops and phones. We'll start to see more use of biometrics such as iris scan, facial recognition and fingerprint scan to replace existing PIN number and password technologies in the coming few years. ATMs will use fingerprint access or iris scan instead of the classic PIN. The call centre will use either voice recognition to verify our identity or ask us to swipe our fingerprint into our phone for access to third-party transactions that currently require two-factor authentication. Automated branches could even use RFID chips and then verify our identity through facial recognition software.

Visa has been working on integrating biometrics into the credit card experience. One prototype propositioned as the Visa e-card incorporates NFC smart-card technology, fingerprint authentication and WiFi negotiation to communicate with both the POS (point-of-sale) terminal and the bank. It was designed by Jacob Palmborg in collaboration with the Yanko design team.

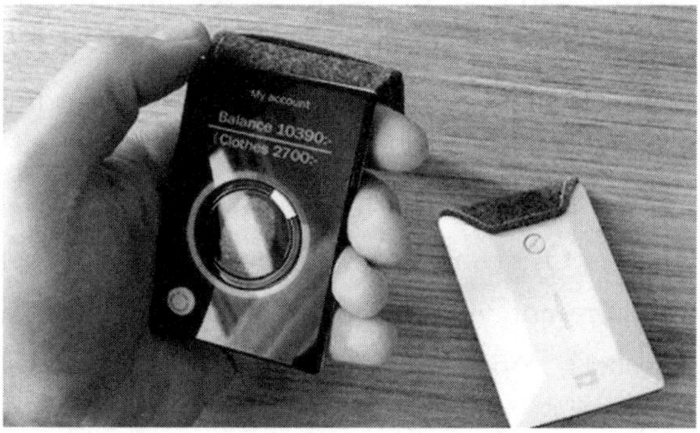

Figure 12.2 Your new Visa card?
(Credit: Yanko Design/ Jacob Palmborg)

If you take a look at the e-card prototype, it looks quite similar to two popular devices these days, namely the iPhone and the NexusOne. It is not hard to imagine iPhone multitouch capability incorporating a fingerprint biometrics reader, or something similar into their fourth or fifth generation of phone. Why sign when you can give the retailer the finger? Sorry, your fingerprint.

Virtual Currencies

A payment system that almost brought down the yuan?

"Hey friend, could you spare a QQ coin or a Linden dollar?" There's been a great deal of debate recently about the revaluation of the yuan/renminbi. But did you know that perhaps the greatest risk to the yuan in terms of competing currency in the last two to three years actually came from a local online currency know as QQ coins.

You will recall we mentioned the astonishing statistic that QQ Messenger is actually the most widely used instant messenger programme on the planet, despite the fact that it is only mostly used in China. QQ Messenger's creator Tencent, however, is also famous for the virtual online currency it created for paying for goods through cyberspace. The online currency, QQ coins, was used for everything from paying for avatars, downloading mobile ringtones, to even online gaming.

This illustrates the real value of online currency and payment systems, something that cash focused traditional bankers have been critical of due to the absence of regulatory control. While some regulators such as the HKMA (Hong Kong Monetary Authority) have taken action to regulate electronic payment solutions such as the Octopus smart-card, particularly in respect of deposit taking, there are precious little rules regarding virtual currency in cyberspace.

Let's take a possible scenario. Your boss decides to pay part of your salary in QQ coins or Linden dollars. Could the tax man come after your virtual revenue if you only spent it online and didn't exchange it for real currency? Oops ... don't tell the IRS I gave you that idea.

This from the *Asia Times*:

> "The so-called 'QQ' coin—issued by Tencent, China's largest instant-messaging service provider—has become so popular that the country's central bank is worried that it could affect the value of the yuan. Li Chao, spokesman and director of the General Office of the People's Bank of China, has expressed his concern in the Chinese media and announced that the central bank will draft regulations next year governing virtual transactions. Public prosecutor Yang Tao issued this warning: 'The QQ coin is challenging the status of the renminbi [yuan] as the only legitimate currency in China.' "[1]

QQ currency speculators have even opened up a Forex trade in the currency, as they have with Linden dollars that power purchases in Second Life. In China, the players have gone one step further, with online vendors hiring professionals to play online games earning QQ coins as currency. Some even use hackers and other methods to steal the coins. They then sell the virtual currency below its official value, at a rate of 0.4–0.8 yuan per coin. The Chinese government initially tried placing capital controls on QQ coins, but that just led to scarcity, driving up their real world value by 70 per cent in a matter of weeks. Considering the QQ Instant Messenger platform has over 900 million subscribers, can this phenomenon be stopped?

Virtual economies are becoming increasingly important, says Wharton Legal Studies Professor Dan Hunter, adding that they could redefine the concept of work, help test economic theories and contribute to the gross domestic product. "Increasingly, these virtual economies are leading to real money trades," notes Hunter, one of a handful of academics closely following this trend.

Edward Castronova, a professor at Indiana University, estimates virtual economies to have a total value of somewhere in the order of $1 billion.[2] Now virtual exchanges are launching to capture trade and speculate on these virtual currencies. The Linden dollar currently trades at between L$276–$300 per US dollar on the Second Life Exchange platform; the Linden dollar even has its own currency abbreviation "SLL". Daily trading volume is a staggering US$6.6 million.[3]

Should we add Linden dollars and QQ coins to your electronic Forex trading platform? It won't be long…

Mobile P2P Payments

As we identified in Chapter 6, remittance payments via mobile phone are experiencing explosive growth. According to the World Bank, today there are 191 million migrants sending over US$270 billion annually, with the G8 markets accounting for 46 per cent of global remittance financial flows.[4] The 20 million people who constitute the Indian diaspora, a group with as much ethnic and cultural diversity as the population it left behind, are spread over 135 countries. In 2007, they sent back to India almost $15 billion, a source of foreign exchange that exceeds revenues generated by the country's highly regarded software industry.

In the past, foreign workers and their remittances received scant policy attention. That's not the case anymore. Immigration has become a major domestic and foreign policy issue from Paris to Manila to Mexico City. Many developing countries, such as the Philippines and Mexico, which have come to depend on remittances as a vital source of external finance, are pushing a pro-migration agenda.

As a result of in-country experience with SMS top-off transfers and "sachet" micropayments, both Smart and Globe, the Philippines' leading operators, have introduced commercial means to have more informal processes around remittances via SMS-powered micropayments sent to anywhere in the world. Using an m-commerce platform that allows for money transfers and micropayment transactions via mobile network operators, users can send money and pay for goods and services such as utility bills, tuition fees, donations to charities, and even buy airtime using local services with offshore operators. Globe's service is GCASH and Smart's, Padala. With close to 70 per cent of the Philippine's overseas foreign workers being women, these P2P payment options ensure that their hard earned income is directed exactly to the intended destinations—normally their family and friends.

This $1 trillion industry and opportunity has been largely missed by banks which decided that migrant workers were too low margin to be an attractive customer. This was the case in Kenya where the four big banks have 3.5 million customers between them, but M-PESA, the mobile payments solution, already has 11 million customers, and that's in only three years.

More traditional banks have jumped into the fray in recent times. Mercantile Bank of Michigan announced in November 2009 that they had launched a mobile payments solution for their customers, incorporating technology from S1 and utilising PayPal.

> "This deal reflects the growing functionality of the mobile channel and is a strong signal to banks of where customer expectations are headed ... Given the near-ubiquity of mobile devices and fast adoption of smart-phones in particular, we can expect an increasing convergence of trusted banking relationships, personal payments and mobile."
>
> Bob Egan, Global Head of Research and Chief Analyst at TowerGroup, a market research firm based in Needham, MA[5]

The payment solution allows Mercantile Bank customers to send money via the PayPal network to any individual who has an email address or mobile phone number. If they don't have a PayPal account, they will have to create one before they can receive the funds.

CashEdge has released a service they call **POPMoney,** which is already subscribed to by First Hawaiian Bank and PNC Bank based in Pittsburgh. POPMoney extends the P2P payment model/service to allow bank-to-bank payments. From within either online banking or a mobile banking app, customers can send money to a recipient by using their email address, mobile phone number or their bank account details. If you are a customer of a subscribing bank receiving a payment, then the payment goes straight into your nominated bank account. If not, you simply go to the POPMoney website, register and give the bank account details where you would like it deposited. POPMoney uses the SWIFT network to finalise the transaction for you.

One of Twitter's founders, Jack Dorsey, has just launched **Square.** It's yet another company that attempts to tackle the issue of making mobile payments work, but instead of trying to become a completely new payment processor, they are depending on the existing infrastructure of plastic credit cards built by Visa, MasterCard and American Express.

Retailers, service providers or merchants that want to accept payments via Square need to purchase a small magnetic card reader that they can plug into the audio jack of their mobile device. Currently only the iPhone is supported, but there are plans to release an Android and BlackBerry version. You swipe your credit card across this magnetic card reader, the data is translated into an audio signal, and finally that sound is fed into a payment system just like a traditional credit card payment.

Figure 12.3 The Square payment device (Credit: Squareup.com)

Verifone PAYware has jumped into the same space as Square, practically within days of the Square announcement. Their solution looks a little clunkier than that of Square, but then again Verifone is at least a much more experienced operator in the POS (Point-of-Sale) space and could argue they would have fewer execution issues rolling out their technology.

My prediction is that in 2010-11, banks will be falling all over themselves to launch mobile applications with mobile payment options, as will Visa and MasterCard. By 2011, when NFC (Near Field Communication) enabled handsets are the norm, I expect the POS infrastructure to start to catch up finally to the promise of true mobile payments. Orange/3 in the UK has already launched contactless credit cards with Barclays, and is offering support for this through NFC SIM cards.[6] How P2P payments are done at that point remains to be seen. I am betting PayPal will remain the dominant platform as part of this equation.

Some start-ups are even using the Twitter stream and PayPal for micropayments. **Twipper** and **Twitpay** (probably not a big winner in the UK market, I would guess) are two such services. Payments are sent by tweeting a simple message to the service, directly into their Twitter stream. When the message is tweeted by the payer, it is processed to identify the amount, currency and recipient, who is then notified of the transaction. Both services use PayPal as the mechanism to complete the transfer of funds between users.

Point-of-Sale and Credit Card Evolution

As we focus more on engaging customers at the point-of-impact (see Chapter 13), POS technology must evolve to support a more interactive sales experience. At the moment, POS units are dramatically underutilised from a customer experience perspective. I've never seen my name, an offer, or any other personalised information appear on a POS terminal. Even at Starbucks, when they swipe my Starbucks card, it doesn't say, "Hi Mr. King, welcome back for your tall, skinny, no-whip hot Mocha…" It would be pretty cool if it did that, don't you think?

We'll talk more about the mechanics of utilising intelligent POS terminals for service selling in the next chapter. The technical side of the requirement, however, is that these POS terminals need to integrate with RFID chip technology, talk to app phones as they come near, either by displaying a message or sending a broadcast message to the phone, plus they need to integrate in real time with the bank or retailer's CRM system and give you that feedback. As for the checkout experience, the days of waiting in line to make a POS transaction may be numbered.

"Today, if you used RFID in its purest form, you could walk into a store, load your cart and walk out without talking to anybody, because they would know who you are."

Randy Carr, Vice-President of Marketing,
Shift4, a developer of enterprise payment solutions[7]

Whatever replaces the legacy, the POS terminal will undoubtedly be based on mobile technology, and will likely use RFID, Internet protocol or both. Ironically, the US is probably going to be last in line for these changes. Why? Largely because of the huge investment already made in legacy POS telephony and equipment. The shift to IP and cloud-based technology in POS is already on the way, but both American merchants and card providers are reluctant to make the shift from magnetic card and dial-up systems. Thus, consumers in the US will probably be among the last to make the switch to the new generation of technologies emerging for retail payments.

> *"The way you change that is a new terminal. That is a surmountable hurdle. It's not going to happen next year, but it could happen in the next seven to eight years."*
>
> George Peabody, Director of Emerging Technologies Advisory Services, Mercator Advisory Group[8]

Phase I: Chip and PIN (current evolution)

Smart-card based credit cards, otherwise known as Chip and PIN, are based on the EMV standard for interoperation of IC cards or "Chip cards" and IC capable POS terminals and ATMs. The name **EMV** comes from the initial letters of Europay, MasterCard and VISA, the three companies that originally cooperated to develop the standard. Europay was absorbed into MasterCard in 2002, and in 2004 JCB (formerly Japan Credit Bureau) joined the organisation. American Express finally relented and joined EMV in February 2009.

In the UK, Chip and PIN was initially trialled in Northampton, and was rolled out nationwide in 2004 with advertisements in the press and national television touting the "Safety in Numbers" slogan. As of 1 January 2005, the liability for signature-based transactions in the UK was shifted to the retailer. This was designed to act as an incentive for retailers to upgrade their POS systems.

New cards featuring both magnetic stripes and chips are now issued by all major banks in the UK, and have become the norm outside of the UK, largely in the last few years. The University of Cambridge security group and others have been critical of Chip and PIN technologies citing fraud opportunities. Given the higher incidents of fraud with traditional magnetic stripe technology, however, Chip and PIN is generally seen as a significant improvement. Cardholders who are incapable of entering a PIN because of a mental or physical disability can contact their bank to be issued with a Chip and signature card.

The US now appears to be the odd one out as Canada, Mexico and most of Europe and the developed economies of Asia are already rolling out the standard. If Brazil, Mexico and Turkey are adopting the standard, you'd think it was time for US card companies to fall into line.

Figure 12.4 Visa Classic credit card with integrated chip
(Credit: Visa/DBS Bank)

The issue in the US is the legacy POS infrastructure. Without the regulator mandating a new standard, it is unlikely that US cardholders will see the new technology any time soon. Thus, many American travellers abroad are now finding their plastic useless at retailers, transit stations and ATMs. It will take some consumer lobbying to make this happen in the US, I think. Until then, some US consumers have taken to travelling with two cards—their mag-stripe card and an upgraded smart-card. This disparity has prompted such stellar blog and media headlines as:

- **"American credit card users are cavemen in a chip-and-PIN world."**
- **"US credit cards becoming outdated, less usable abroad."**
- **"US magnetic stripe credit cards on brink of extinction?"**

The US has been pushing a possible contactless standard as a replacement for magnetic card. But with the rest of the world already rapidly adopting Chip and PIN, it looks like the US will have to admit defeat grudgingly and start the roll-out of new POS technology sooner or later. Either that, or the US market will need to skip the whole Chip and PIN phase and move on to Phase II.

Phase II: App phone integration (2 to 4 years)

Workarounds for app phone integration with credit cards and POS systems are already underway. The issue with the integration of these with POS terminals is twofold. Firstly, how does the app phone communicate with the POS terminal, and secondly, how does the consumer authenticate or provide proof of identity?

There are two current technologies in place that would enable app phone payment. The first is NFC discussed in Chapter 6 on mobile banking, which means that the phone contains the IC (Integrated Chip) or smart-card that enables them to be used identically to the Chip and PIN credit card. However, to set the cat among the pigeons, as one might say, it looks very much as if Apple is planning NFC integration into its future iPhones. In December 2007, Apple filed a patent with the US Patents office describing a method of integrating RFID circuitry into a touch sensor panel.

> "The RFID antenna can be placed in the touch sensor panel, such that the touch sensor panel can now additionally function as an RFID transponder. No separate space-consuming RFID antenna is necessary. Loops (single or multiple) forming the loop antenna of the RFID circuit (for either reader or tag applications) can be formed from metal on the same layer as metal traces formed in the borders of a substrate..."
>
> Abstract from Apple's patent application on RFID integration[9]

Based on the iPhone's massive popularity, this would probably end the debate on whether NFC-enabled handsets were going to be ubiquitous or not in short order. My hope is that this becomes a reality late in 2010, and if so, physical credit cards will be virtually redundant by 2012. This is very exciting for consumers, but terrifying for banks and card companies. One wonders whether regulators in the US might move to stop Apple to protect the interests of the card companies.

Ironically, the second viable technology utilising either an application IP-based or call-based solution is workable right now today—without any development of a supporting platform. There are already providers in the market that supply secure authentication utilising both methods without even the need for a POS terminal at all. We have discussed Square, PAYware and others already.

One solution that is already well established as an application-based payment system is that from a four-year-old company based in Queensland, Australia. The company, QPay, already has deals with the Indonesian government, Singapore Yellow Pages, Westpac Bank and many others. QPay requires only a simple, one-time registration and authentication for new users—through a hybrid voice and voice-to-text based system—where they lodge the details of their debit or credit card. After registration, new users can effect payments to whomever they like by using just their mobile and their PIN code. Card details don't even have to be supplied for an individual transaction because these are stored securely with the bank handling the back-end authentication in tandem with QPay.

> "What we do is mask sensitive data, along with addressing the flaws of voice biometrics and tokens, and we're cheaper than a normal credit card process for merchants. We're Westpac's first payment aggregator, and their first globally approved payment aggregator for Visa and Mastercard. Plus we're platform-agnostic, telco-agnostic, network-agnostic and phone agnostic..."
>
> Greg Walter, CEO of QPay[10]

When a consumer sees an advertisement that tempts them, they can simply send a product code from the ad via text message to QPay. An automated call back is used to confirm the purchase, then the funds are transferred to the retailer and the product is shipped. Or, you could call up a tradesperson to come and fix your plumbing at home. You could then enable a simple payment to this individual by getting him to register with QPay and transferring the funds securely to his account. The advantage for merchants and retailers is that there is no POS equipment required, no registration fee, and no infrastructure costs.

Which of these two methods—NFC or RFID phone-based, or

application/call-based—will come out on top? My guess is that it will be a combination of the two, but over time the simplicity of NFC will win out for real-time interactions at the retailer's store or at the train station, for example, whereas application technology will work for virtual stores. Already Visa and Barclay are trialling the one-pulse NFC system integrated into mobile phones. If Apple or Google adopts the same, then we can expect this method to dominate.

QR/Semacodes and other such methods could also be used, as could a Google Goggles type technology with your camera in your app phone, where you take an image of an advertisement you see on a billboard and you have the option of purchasing that item or product through the mobile Internet.

Phase III: Card-less, phone-less, personalised?

In *Minority Report*, we see Tom Cruise's character interacting with advertisements that speak to him about his last purchases using only retinal scans to identify the customer. While the futuristic technology from the movie may be decades away, the likelihood of some adaptation of this technology making its way into the retail experience is not that far-fetched.

POS terminals that use directed audio or send alerts via location-based messaging is highly possible. In fact, existing RFID technology could effectively enable you theoretically to walk in, select the items you want at a store and walk out again, without having to specifically engage a cashier at all—the charges would just be automatically made. You could then receive an instant receipt on your mobile phone.

While Phase I and II deal largely with the evolution of the payment device (mag-card, IC card or app phone), Phase III requires an evolution in the POS environment. Once again, cloud computing could be the vehicle that breaks this platform right open for greater collaboration and more interesting application. Today, we are restricted somewhat by legacy POS infrastructure. The most likely short-term improvement is that you could have roaming advisors or cashiers in the store with a hand-held payment terminal that can process your payments in real time.

Conclusion—Mobile Payments and Quickly

The conclusions are inevitable. If you think Internet banking take-up was rapid, wait till you get a load of mobile payments. Banks and credit card companies are used to owning the infrastructure for payments processing. However, what we are seeing is a deleveraging of the retail payment experience from the back-end banking system. That is, banks are simply no longer going to be necessary when it comes to point-of-sale, neither are credit card companies.

Do you remember payment by instalment, layaway or lay-bys? My kids have never heard of them, and I can't remember seeing them for at least a decade, but they used to be a popular method of getting through the Christmas credit crunch in the old days. But layaways have mostly disappeared because credit cards were a better idea. Just like layaways under threat from a new payment mechanism, physical credit cards may also be on their way out due to mobile payments.

The other poignant issue is that credit cards are a risky business for banks in the post-global financial crisis world. So a shift to debit cards is good business because it promotes less consumer debt and is a risk adverse strategy for banks. It does take a rather sizeable chunk out of retail earnings though. Debit card usage is set to rise considerably over the coming years, making up almost 50 per cent of all transactions by 2015. As the debit card merges with your app phone and POS systems allow you to use your phone to pay, it will be impossible to continue to support physical plastic cards. It might even be good for the credit ratings of our debt-laden economies.

As cloud computing and new IP-enabled retail POS devices allow for more and more retailers to accept payments from your app phone without having to plug into Visa, American Express or MasterCard, who will be in control? The consumer and the retailer! There simply is no real value provided by SWIFT, Visa, MasterCard and others when P2P payments and cloud computing-enabled POS systems hit our stores.

As long as these new systems or solutions provide expedient means for payment, are secure and yet flexible, then as consumers, we will adapt. What could be more convenient than waving your iPhone over a contactless pad at a retail outlet, ordering your pizza through your MoBank-enabled mobile phone, or squirting some cash through the ether to your local gardener who just mowed your lawn? Especially if you don't need to carry extra plastic, jump through hoops to qualify, or pay 27 per cent interest per annum.

If banks, merchants and card companies don't move very, very fast, they will find themselves out of the loop on this one.

Endnotes

1   *Asia Times Online*. www.atimes.com, 5 December 2006

2   cnn.com/technology. "Virtual currencies' power social networks, online games", John D. Sutter,19 May 2009. http://edition.cnn.com/2009/TECH/05/19/online.currency/index.html

3   LindenX™ Market Data, SecondLife.com. http://secondlife.com/statistics/economy-market.php

4   Migration and Development Brief 8 – World Bank. worldbank.org

5   Mercantile Bank press release, 3 November 2009

6   "Barclays and Orange unwrap contactless credit card", Silicon.com, 7 January 2010

7   "What will credit cards look like in 25, 50 or 100 years?", Jay MacDonald. CreditCards.com, 17 February 2009

8   ibid

9   US Patents & Trademark Office, Serial-No 965560 (TOUCH SCREEN RFID TAG READER, Apple Inc.)

10   "QPay sets sites high with Westpac, GYP wins", Petroc Wilton Communications Day, 4 December 2009

13 Death of Advertising—Predictive and Precognitive Sales and Marketing

Going, Going, Gone...

THIS shouldn't be a surprise, but for many of the traditional marketers in our midst, the first reaction to the data and analysis I will present in this chapter could be scepticism or outright denial. If you Google these statistics, you'll find they are all too real.

The fact is that in most banks today, the marketing department is completely misaligned to the reality of the marketplace. The skills are wrong, the methods are wrong, and the people are wrong. It is the one department that needs to be dramatically re-engineered with immediate prejudice. Advertisers still talk about "media" and "interactive" or "digital" like they are two separate advertising universes. Marketers talk about "offline" and "online" as if never might the two be equal. These are 20th-century classifications in a 21st century **BANK** 2.0 world.

In 2008, the Internet surpassed all media except television as the primary source for national and international news.[1] This has taken its toll. In March 2009, the 146-year-old newspaper *Seattle Post-Intelligencer* or the "PI" as it was known, closed down, citing rising costs, falling revenues and declining circulation. Since January 2008, at last count, 53 regional newspapers in Britain have folded. Of the top 25 newspapers in the US in 1990 (the year newspaper employment peaked), 20 of those newspapers have seen declines (on average reporting circulation down by more than 30 per cent), and two have been closed down or declared bankrupt. *New York Times* reported a 30 per cent fall in advertising revenue, resulting in a $35.6 million loss for the 2009 third quarter alone. In November 2009, the *Washington Post* announced the closure of all its remaining US bureau offices (New York, LA and Chicago) outside of the capital, in addition to Miami, Denver and Austin closures in recent years.

A lot of people are asking the question, "How do we save the newspapers?" Unfortunately for those asking that question, it's not just newspapers we have to worry about.

In 2009, TV advertising revenues in Australia fell by more than 12.6 per cent in the first half of the year. In the first quarter of 2009, the US recorded losses of more than 14 per cent in TV ad revenues in normally stable locations such as the Bay Area and New York, and this is expected to suffer a total decline of 22 per cent for the year. Declines of 27 per cent and more were recorded in radio ad spend for the US for the first half, even worse than the decline in TV commercials. In the UK, TV ad revenues were down 12–14 per cent in 2009.

While many attribute the further decline of traditional media to the global financial crisis, it doesn't explain the trending over the last five to six years since new media started to bite. In fact, a report in December 2009 in the UK was heralded as good news when it projected ad spend in 2010 would only decline by 0.2 per cent! Yes, the industry is now celebrating when traditional ad spending doesn't decline in any given year. A recent report commissioned by OFCOM (Office of Communications) forecast the value of TV ads in the UK could fall from £3.16 billion in 2007 to just £520 million in 12 years' time. That's an 83 per cent decline.

The erosion of *offline* media and the increase in Internet usage, combined with a blending of news, entertainment and information sources, mean that all forms of media are in a constant state of reinvention in the **BANK** 2.0 world. More and more consumers are choosing what messages they want to listen to, and when. Technology is assisting them in these choices. In 2003, stable brands such as Coke, Procter & Gamble, General Motors and others all started to understand the impact of the TiVo effect on the TVC market. That year was the start of big budget cuts on TVCs and the move to alternative media.

The main culprit for the 2009 US TV and radio ad spending decline was the imploding auto industry, which traditionally provided around 20–35 per cent of the total advertising revenue for broadcasters, but plunged 30.9 per cent in ad spend in the first three quarters.[2] The financial services sector is traditionally the second biggest spender on these types of ads, but they have cut heavily too as the recession has taken effect and as marketers realise that they are not getting the bang for their buck that they used to get. It's unlikely that this revenue will ever come back.

In September 2009, the Internet Advertising Bureau announced that the UK has become the first EU economy to see advertisers spend more on Internet advertising than on television advertising, with a record £1.75 billion online spend in the first six months of the year. The milestone marks a watershed for the embattled British TV industry, the leading ad medium in the UK for almost half a century. In the US, Internet ad spend had already overtaken total TV spend as the leading choice for media buy in 2007.

Media lines are blurring, and as we've discussed, consumer behaviour is in a state of continual change. The merging of information sources is having an impact on the way in which technologies are developing in response to changing consumer behaviour.

Newspapers have been the form of traditional media hardest hit by the Internet as readers continue to abandon print in favour of online news portals or search and aggregation destinations which increasingly act as a window on the world. While some magazines are supported strongly by an online presence, as yet the world of e-books does not represent a threat to physical books (or you'd be reading this on your Kindle right now). As e-paper and such technologies start to come of age, this could change more quickly. As of today, we see Rupert Murdoch and other newspaper moguls trying to build a business strategy to combat this, which will require users to pay for access to online newspapers (See PaidContent.org—"Murdoch Paper Blocks UK Aggregator Before Paywall Goes Up").[3]

"Digital Natives", the Y-Gen and younger users of technology are spending greater time online more quickly than their predecessors, and they are changing the way in which they consume and access news and entertainment through a myriad of devices. They don't think of TV or newspapers as a traditional source of content; they think of the browser and mobile devices as "traditional" media. They don't think of offline versus online; they think of content, downloads, media streams, P2P, blogs and networks.

The prosumer is showing his face here too. In 2008, 58 per cent of newspaper websites featured some form of user-generated content,[4] more than double the previous year. Online news sites now include a mix of user-generated photos (58 per cent), home videos (18 per cent) and independent articles (15 per cent). Meanwhile, the number of newspaper sites that are allowing readers to comment on articles has more than doubled to 75 per cent. Universal McCann recently identified "consumer publishing and the rise of consumer power as the biggest trend in marketing communications today". Shocking! Yes, Universal McCann—this is the prosumer and they don't value traditional media over "interactive". Live with it.

Today, mass media is **social,** and journalists, consumers and bloggers alike can all have a say. It makes life tough for journalists, and it has the ability to skew news or content. Don't believe everything you find on the Internet.

Advertisers and marketers alike are having a hard time dealing with these changes, and it's not just in advertising. In banking, it has long been held that direct mail campaigns are one of the most successful methods of acquiring new customers. Well, those days are over too. Direct mail offerings have been declining rapidly since 2006. In 2009, fewer direct mail was sent by banks than in the year 2000. Direct mail has declined 32 per cent since 2007, and is expected to decline a further 38 per cent over the next few years.

BIA/Kelsey, a financial advisor to US-based media companies, estimates that TV station ad revenues will rise 3 per cent in 2010—or $500 million—to $16.1 billion. BIA estimates that by 2013, total station ad revenues will only inch forward to $16.4 billion. This is down to 1990 levels of ad spend for TV in the US. The media group now says 2009 will end at $15.6 billion, down 22.4 per cent from last year's $20.1 billion mark. TV stations reached an all-time ad revenue record in 2006, when it was at $22.8 billion.

What about the not-so-new media?

Now to be fair, Internet advertising did take a dip in the first half of 2009. How did it compare with the likes of TV (12.5–19 per cent), newspapers (30+ per cent) and radio (14.7–29 per cent) in developed economies?

It suffered a 5.4 per cent decline against 2008 figures (Figure 13.1). 2009 will be the first year that Internet advertising has suffered a decline in over six years straight of increased spend. In comparison, TV ad spend and direct mail have been in decline since 2006, newspapers since 1990. Radio ad spending has been virtually flat since 2006, hovering around the $20–$21 billion mark annually.

The only traditional medium for ad spend that has actually increased in recent years is outdoor advertising. Why? Well it has gone digital. Video screen advertising outdoors has saved this niche market from an otherwise ignominious demise.

Total Internet ad spend will likely reach around $22.6 billion in 2009, compared with $23.4 billion in 2008. How does that compare with TVCs historically? The highest ever annual TVC spend in the US was in 2006 at $22.8 billion. So even in a year with the biggest recession since 1929, Internet ad spend is still tracking at the same level as the best year on record for local TV ad spend ever!

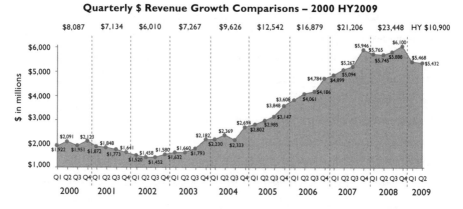

Figure 13.1 Internet advertising revenue 2000–09
(Source: PwC/IAB Internet Advertising Revenue Report)

What was the mix of ad spend through the online medium in 2009? Search continues to lead, followed by display banners and classifieds— Search revenue accounted for 47 per cent of 2009 second-quarter revenues, up from the 44 per cent reported in the second quarter of 2008. Display advertising, the second largest format, accounted for 35 per cent, followed by Classifieds (10 per cent), and lead generation (7 per cent) of 2009 second-quarter revenues.

> "While the overall advertising market has continued to be impacted by current economic conditions, marketers are allocating more of their dollars to digital media for its accountability and because consumers are spending more of their leisure time online."
>
> David Silverman, partner, PricewaterhouseCoopers LLP IAB
> Internet Advertising Revenue Report, 2009

And the new, new media?

Pretty much every media-savvy advertiser recognises the new opportunities presented by mobile phone marketing. Mobile phone marketing will account for 11.7 per cent of total digital advertising spend by 2014, it has been claimed. According to new projections from industry analyst Berg Insight, the mobile marketing sector will be worth an estimated €8.7 billion (£7.8 billion) in the next five years.

Despite massive adoption rates, new and improved app phone devices and increasing application of these devices for more diverse use, advertising and marketing dollars flowing to mobile lag well behind consumer usage of the channel. The long-term growth trends, however, are stellar. Spending on mobile advertising is set to increase rapidly over the next five years, with growth of close to 500 per cent between 2008 and 2013, according to eMarketer, which has estimated that mobile ad spending, including messaging-based formats, will increase to US$1.56 billion by 2013.

There is a lot of disagreement in this space because it remains largely untested. Estimates for ad spend span a broad range. Yankee Group predicts $184 million in 2009 spent on the mobile platform, while the Mobile Marketing Association forecasts spending will reach $1.7 billion in

2009. They can't both be right. This disparity is indicative of the relative immaturity of the channel.

If used with strong targeting and customer analytics, mobile will be able to revolutionise direct marketing. But beware—spam is still spam. Don't fall into the trap of broadcast or shotgun SMS campaigns to your wider database.

In any case, mobile marketing is sure to be the next area of focus for new media advertisers. Google is set to focus on integrating more mobile marketing into their devices both through applications and augmented reality. Apple certainly believes in the future of this model. It was announced in January 2010 that Apple had acquired mobile advertising company Quattro Wireless. Quattro, which places ads on mobile Web sites and applications, was reportedly acquired by Apple for a decent $275 million according to *The Wall Street Journal.*

Banner blindness, SPAM and TiVo

Traditional advertising has been about eyeballs—the more eyeballs you reach and the more frequently, the more chance you have of brand recall when the decision to purchase is being made. In the early 1900s, consumers were faced with few choices of brands and advertising was limited to mostly in-store promotion. By the mid-30s, consumers faced exposure to hundreds of products a year by virtue of radio. These days, the average consumer is confronted by thousands of advertising images on a daily basis alone. The result is that advertisers are trying to shout louder for our attention, and we are increasingly finding ways to reduce this impact.

Our browsers now have pop-up blockers. Our email has spam filters, and we have tougher privacy laws so advertisers can't exchange our contact information. We have do-not-call lists for outbound telesales. And we have TiVo.

TiVo is a hard-disk-based video recording device that lets you "tape" your favourite shows and play them back later. But TiVo has one other cool feature—it lets you strip out the TV commercials. When TiVo came out in the US, advertisers lobbied Congress to ban the "filter" feature. They failed. Advertising response rates through TVCs crashed. Companies slashed their TV budgets, and advertisers bemoaned the existence of TiVo. How did they finally adapt to the TiVo phenomenon? *Hollywood Reporter* revealed that Coca-Cola, Ford and AT&T forked out over roughly $35 million each for the opportunity to be featured in American Idol, America's most watched TV show in 2008. And have you ever noticed how Simon, Randy and Paula (or Kara these days) are all drinking Coke? This is the advertising industry's attempt to keep us looking at products when it is clear we're not watching TVCs anymore.

In a world where the broadcast style of advertising is increasingly failing to get results, banks will be forced to adopt a much more personalised style of marketing. When customers come to a branch, we will need to know with a high likelihood what they will need so they don't have to wait. If they want to apply for a loan, we will be able to approve it instantly—no forms, salary records and bank statements needed. If they want a mortgage or investment product, we will have the application forms pre-printed with all their details so they only need to sign, or give us their fingerprint using their digital ID card.

The days of direct mail campaigns, newspaper ads and billboards will largely disappear as we dedicate our marketing team to customer intelligence, predictive and permission-based marketing, and better, smaller segmentation methods.

I will make a pretty scary prediction here though. With TV ad spend decreasing so significantly, just like we have seen with newspapers in the last decade, it is inevitable that we will lose many local and regional TV stations. In fact, as the digital natives choose downloads and interactive TV as their primary source of show content, it is possible that free-to-air TV will not be able to survive. If there is no ad revenue, how can they? Product placement will still work as a medium for downloaded content. CNN, CNBC, CBS and the other networks, better start figuring out new revenue models based on downloads pretty fast.

Not Just for Kids With Acne

Ok, so maybe we've convinced you that Web 2.0 and all that is a force to be reckoned with, but you might argue that your customers are mostly seniors who still like going down the branch to have their passbook stamped, and that your most valuable High Net Worth clients simply don't use the Internet for banking … WRONG!

High Net Worth Individuals (HNWI) are statistically amongst the most time-poor, most tech savvy early adopters you will meet. If you don't believe me, just look at the top banking search terms used in November 2009. Guess who is doing the searching with terms such as Private Banking Offshore, High Net Worth, Offshore Banking, Bank Guernsey? I'll give you one guess and it's not the pimply teenage iPod generation. It's your HNWI customers.

The traditional marketing approach through print and direct mail is simply no longer relevant to the best and most profitable customers. HNWIs are highly mobile, global and time-poor.

Table 13.1 Google Top Asia Banking search keywords in November 2009

Bank or Banking	Wealth Management
Banking	Private Banking
Private Banking Offshore	High Net Worth
Offshore Banking	Private Banking Asia
Bank Guernsey	Priority Banking
Business Banking	Credit Suisse Private Banking

According to both Nielsen studies and research from the Pew Group, 60 per cent of adults in the 55–65 age demographic are regular surfers and prefer Internet banking over branch, and over 68 per cent of Internet users who choose Internet banking over branch are in the 30–55 age bracket.

With respect to continuing adoption of new technologies, think about the demographics of BlackBerry users, and how long ago they started using these technologies. Do you think such individuals are going to find it difficult to make the switch to mobile payments technology?

When Push Comes to Shove

Push marketing is ingrained in traditional Marketing 101 theory. The prevailing theory is that you push enough messages down the throat of your target audience, and when they have to select a suitable product they'll choose yours because they are most likely to remember you. Well, these days we are getting pushed so many messages that the noise is unbearable.

One-to-one marketing and targeting key influencers in social marketing settings is your future. Not traditional direct mail or mass market marketing, but marketing controlled by the consumer. Let customers make the decision on what information you can send them. The customer will let you know when they want home equity offers or the opportunity to assess a new health plan offering. There is already the desire to truly manage their own channels. That's why on average we have 10 visits to the Web by active Internet banking users each week and why mobile banking is the fastest growing channel in our current arsenal—if we have the apps.

Call this new commitment **intelligent, non-intrusive, permission marketing.** Give customers an alert configuration engine that allows them not only to dictate what they will see, but also through what channels. Give them the ability to choose how you mine their transactions and previous interactions with the bank and retailers to provide a better service.

Use mobile entertainment devices to offer a new innovative channel. These move beyond children's toys. An investment seminar can be loaded on a memory stick, either in the form of a podcast (MP3 file) or a video file on a memory stick; put it in the mobile device or in a memory stick-enabled cell phone (which does exist), and give it to emerging affluent customers. They may think, "They get technology and what it can do." Is it worth $250 to you to give your customer a keepsake or a cell phone provider credit in order to grab a $250,000 investment? Sounds reasonable, doesn't it?

Send podcasts by your organisation's investment guru to targeted customers. And, for extra measure, send them to a financial blog hosted by your company. Consider new channels such as Internet TV with your own financial channel. Use Twitter to provide awesome customer advocacy experiences which demonstrate the brand's commitment to customers. Integrate instant messaging capability right onto your websites and mobile apps so you can generate leads in real time through the contact centre.

When our customers are out and about, use location-based messaging capability to target them with relevant offers from retailers that they have purchased from before. How can we tell? Use their credit card history and trawl that to create the relevant offer pool preloaded with the network operator's CRM tools.

Guess what. Most banks are doing none of this right now. Our banks are competing against 21st-century players, but stuck in a 1970s marketing paradigm. We must evolve—fast. There is no downside. With 90 per cent of our transactions taking place through self-service channels, how could we imagine that new media such as YouTube, Twitter and other such mediums are going to continue to do anything less than grow. The only real question is, why aren't we already doing this?

Push to pull, point-of-sale to point-of-impact

A bad sales experience can feel very much like you've been used or abused. Persuasive sale is the technical art of selling someone something that they don't necessarily need, but convincing them long enough to get the commitment to sign. Indeed, when presented with enough pressure, customers will often commit just to terminate the sales conversation. Once bitten, twice shy when it comes to this approach. When we get pushed into a buy, we don't usually come back.

A **good sale,** however, feels like the brand has done something right by you. Even better, a good sale can feel like you've been done a great service. In fact, when we get great service and we feel like we came out on top in the sales conversation, or we got exactly the right solution, we normally come back for more because it was a great experience.

Combining this basic tenet with the concept of permission marketing and better customer analytics, we realise that push won't work in the **BANK** 2.0 paradigm. Indeed, we instinctively have to understand our customers so well that we will anticipate their needs. When we get this right, the subsequent offer will not only feel like a good sale to the customer, it will be **great service.** This is why push-based message marketing needs to change to pull-based, point-of-impact, service selling.

Phase I: Distributed applications at point-of-impact

Right now, your bank could be doing a lot more to reach customers at the point-of-impact. The point-of-impact goes beyond simply the point-of-sale; it includes wherever the sales journey might be initiated with a possibility of closure. With Internet technology this is very simple. In future

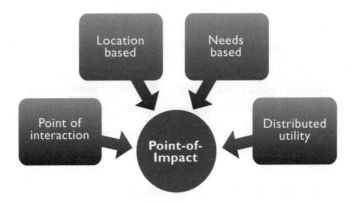

Figure 13.2 Point-of-impact selling strategies

with more integrated wireless capability and improved handset bandwidth, anywhere our customer interacts with a vendor or sale process, your bank can integrate into that experience either from an offer point of view or a payments perspective. From Web to mobile to IP-enabled point-of-sale, distributed application technology will become standard practice in the next five years.

A few such examples of technology capability to get to the user at the point-of-impact would include offering travel insurance within the online booking engine of British Airway's website, offering an instant pre-approved mortgage quotation on a property website, or offering a special car loan proposition on a dealer's website. But speed and targeted placement will be the critical factor in all of this, so we will still need to integrate this with a customer profile—this is not just about third-party banner advertising.

Phase II: Predictive selling or triggered offers

Customer analytics also enable us to predict, anticipate or respond better to opportunities for target clients when these present themselves. This is where the customer dynamics team comes into play. They have to create a list of these target opportunities and possible event triggers, so the offers are ready to go when the event occurs. Without this vital element, these events will just pass us by. Table 13.2 lists some examples of event triggered offers.

Table 13.2 Event triggered offers

Event Type	Event	Event Description
Significant balance change	Investment needs	Customers account holdings increased by large or significant amount. Lead delivered to banker next day who contacts customer with offer e.g. a financial planning appointment.
Large transactions	Withdrawals/ deposits	A transaction out of the ordinary for that customer e.g. greater than average for last three months. Lead delivered to banker next day who contacts customer to identify and fulfil changed needs.
Personal tax loans	Redraw request/ payout request	Call centre receives request for redraw or payout. Lead delivered to banker next day who contacts customer to identify and fulfil changed needs.
Term deposits	Renewals/ upgrades	Maturing lead delivered to relationship manager. RM contacts customer to renew or increase deposit, perhaps offering better rate through a structured product or something similar.
Home loans	Fixed rate rollover	Banker contacts customer to offer various options including home equity draw down, line of credit options, etc.

Think of this approach from a customer point of view. If you were contacted as a result of one of these events, wouldn't you likely perceive that as *great service*. You wouldn't necessarily think of it as a sale process at all.

Phase III: Precognitive selling

As IP-enabled technology continues to progress, point-of-sale equipment will also migrate to the cloud or integrate with our app phone's capabilities, as we've already discussed. In this environment, banks will be able to respond with an offer on presentation of the card/chip/phone for payment.

That offer may include other options for payment, for example, from an offset account, or an offer for a line-of-credit that is at a much lower rate than the credit card, or an interest-free period. The customer presents his fingerprint to verify his identity and accept the offer as payment at the point-of-sale.

Let's illustrate it this way by comparing today's approach versus the precognitive opportunity. See Figures 13.3 and 13.4. We are very close to being able to offer this sort of great service and value to our customers.

Figure 13.3 Retail purchase event – credit card promotion

Figure 13.4 Precognitive event at point-of-impact

Conclusion

Segmentation and customer intelligence through customer analytics capability will be the key. More than simply segmenting customers, we will need to understand how they behave, what they do, how often, and through which channels. Currently, we don't even understand which transactions go through which channel for existing customers. Marketing must understand the why and how, and ask customers what they want. Perhaps we do that already, but we certainly don't use that data effectively to sell.

We also have to deliver marketing messages in very different ways. In fact, we have to reinvent the marketing function. Branding will stay with us as it always has, but our campaign approach needs to be a lot tighter, and needs to be rapidly actionable. Traditional methods of campaign promotion through direct mail, radio, TV, newspaper, classifieds, outdoor billboards and so forth, will need to be abandoned completely in the near term. Some of those mediums may still be used for branding, but not for promotion or campaigns. Why? They are simply losing their effectiveness too quickly to remain a core skill set of the business. We need to move campaign and offer management to more relevant and actionable mediums, namely Web, mobile, social networking and viral.

Your team needs to change NOW! If you haven't figured this out by now, your current marketing team is probably going to find the **BANK** 2.0 revolution a little bit difficult to master. Old habits die hard, and the temptation to send off a brief to an agency to produce some creative for a print campaign might be hard to break.

That's why we need new thinking in the team. Half your team over the next five years needs to be focused entirely on what advertisers fondly call interactive or digital, but the two biggest parts of your innovation in marketing need to fit into digital and social networking.

If you are a marketer in a bank and you find yourself shaking your head at this assertion, then you might need to think about a change in career— preferably before your boss decides to make that change for you. However, if you've looked at the numbers and you've come to the same conclusions, then get on board. There's plenty of scope and room for converts in the point-of-impact discipline. Start by getting yourself a Twitter account and set up a Facebook page (if you haven't already). Then form a revolutionary, strategic fighting force known as a customer dynamics team. You'll be on the frontline of growth.

Endnotes

1   *EIAA Mediascape*, November 2008

2   Neilsen Research

3   http://paidcontent.org/article/419-the-pay-wall-will-be-built-times-blocks-aggregator-newsnow/

4   Jupiter Research, European Media Consumption, October 2008

14 The BANK 2.0 Roadmap

Your Critical Path Checklist For BANK 2.0

Review these key questions, which are logical conclusions to the data presented throughout this book. Your answer to these critical questions will indicate if your organisational strategy needs a rethink for 2010 and beyond.

1. **Did you invest more than twice on remodelling, opening, staffing or marketing branch related activities versus what you invested in Internet, social networking and mobile in 2009?**

   If you answer yes, given that the branch represents such a small fraction of your transactional activity and is at level pegging with the Internet channel with respect to revenue, what is the business case that could substantiate this?

2. **Has more than half your marketing team been in the job more than five to seven years? Or in 2009, have you had more staff working on traditional brand advertising and campaigns versus digital, social networking and market-of-one efforts?**

   If you answer yes, given that Internet, mobile and social networking are such a prominent part of the day-to-day lives of every target segment of the retail bank, anything less than a 50/50 split in skill set and budget allocation doesn't add up to the right mix for your brand.

3. **Does the head of branch distribution hold a more senior position than that of the head of Internet?**

   If you answer yes, given the mix of business in the bank today, and given the increasing importance of the Internet and mobile channels, why wouldn't you have these critical leaders closer to the head of retail or the CEO so that bank strategy can be better informed?

4. **Do you have an innovation team that has the authority, protected from veto by the channel and distribution leadership, to launch and manage proof-of-concept customer initiatives across the bank?**

   If you answer no, then take another look at Google's or Apple's innovation strategy and then have a rethink about why your brand and managers have so much trouble adapting to change.

5. **Did your marketing budget for traditional media exceed that of so-called interactive or digital media in 2009?**

   If yes, then you need to go back and read Chapter 12 again, and replace your head of marketing today.

6. **Do you have a working plan as to how you are going to phase out cheques with retail customers?**

   If you answer no, then put the team on to it now.

7. **Do you prevent customer facing staff from using Facebook and Twitter during work hours?**

   If yes, you might like to take a leaf out of Dell's book and revisit this. Given that this is free market research and enables you to convert quickly possible customer service disasters into customer service wins at light speed, what is your excuse for allowing this opportunity to slip past your customer advocacy group?

8. **Are your CEO, head of marketing and head of customer advocacy on Twitter?**
If no, then who is telling them what customers are tweeting about your bank?

9. **Can your customers make P2Ppayments, regular transfers and pay bills from their app phone or mobile phone today?**
If no, then this has just become the highest priority project for your retail bank this year.

10. **Can you approve a personal loan application for an existing customer with a salaried account in real time, instantly?**
If no, then you are probably missing two key components in your basic systems infrastructure today— STP (Straight-Thru Processing), and automated credit risk scoring and assessment.

11. **Do you know what percentage of your visitors to your bank's homepage click on "login" versus other sections of the website?**
If no, then how do you know what part of your website budget should be focused on converting existing customer to cross-sell opportunities?

12. **Do you know which products are the most popular in your market when it comes to new Internet revenue? Or do you know which products you sell more of online than through the branch today?**
If no, then what about what products that sell better through the call centre or through the ATM, rather than via branch or direct selling. I'm not talking about monthly MIS reports or a report your channel guys can compile over the

next 10 days. I'm talking about you tracking these day to day, quarter by quarter.

I could put a bunch more questions up here, but I'm afraid this would make a mockery out of the process. These are serious questions that lead to the sort of reflections the CEO and head of retail of the bank need to be asking today. Only one of these questions relates to IT or technology specifically—STP—but the others all relate to reaching and servicing customers more appropriately in the **BANK** 2.0 paradigm. Let's discuss the possible action plan or "roadmap" for the next five years.

The Future Starts Here

The **BANK** 2.0 roadmap outlines how to tackle the changes required in the core arenas, which are Platform, Channel and Distribution, Customer Intelligence, Marketing and Metrics. These changes are grouped into three areas:

- Technology and Innovation.
- Organisational Impact.
- The Projects Roadmap.

Technology and Innovation

Platform and infrastructure

There are three core changes that will occur within the technology platform of leading banks over the next five years. One change will be the implementation of STP capability, and the other will be the move towards channel-agnostic content and processing capability backed by a dynamic, IP-based service-oriented architecture.

Straight-Thru Processing is a concept and process encapsulating technology that relies on a number of core technologies, including customer analytics and metadata management, consolidating the messaging platform between channels and legacy systems (middleware), credit risk management, settlements, and other core components. STP automates the assessment of risk for low to medium involvement products, ranging from

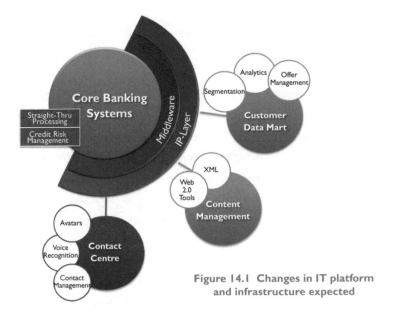

Figure 14.1 Changes in IT platform and infrastructure expected

credit card applications and personal loans, to refinancing of mortgages and lines of credit, and the establishment of overdraft facilities. STP will enable the bank to give instant approval, streamline the fulfilment process and reduce paper handling to an absolute minimum. For existing customers, this will represent a massive improvement in service as the bank will be able to offer instantly a product that a competitor will still need to assess in a more conventional manner. STP will be supported by **Credit Risk Management** and an automated risk assessment platform.

As ATMs, voice, data, branch automation, the Internet and mobile all move on to IP-based operating platforms, the bank's service-oriented architecture will become the DNA of the business. We will continue to migrate business processes to **application server technology** or even into the **cloud,** thus continuing to move away from legacy core banking systems wherever possible. This is done to improve time-to-market for customer solutions and improve adaptability to competitive market pressure, along with reducing capital expenditure. **Autonomic technology** with neural nets will learn new processes that are created by customer interactions. **Content management** with Web 2.0 and XML-based technology to distribute content across multiple channels will generate richer customer experiences and spread the workload of content creation across the business and beyond.

Social networking integration will allow the bank to let customers also integrate content into the banking experience, giving ratings on products and giving service feedback. Banks today would be absolutely terrified of this concept because they don't want free forum feedback that other customers can see. But guess what? Twitter, email, and other such mechanisms already make that possible today. So why not try and prepare for this and integrate it into the bank's DNA.

Additional core components are required to task up the architecture to cope with customer service demands. **Contact management** software will support improvements in customer service by allowing any bank officer or customer representative to see if there are any outstanding customer service issues and to respond to issues that could otherwise affect customer brand experience. It is essential that we have the ability to track contact from customers across every channel, otherwise the service experience as the customer moves from one channel to another will always be compromised.

In-branch, contact centre and relationship management will be assisted through **customer dashboards** which aggregate customer footprint and operationalise credit and offer management. Through the use of customer dashboards, typical service requests will be streamlined so that in-branch experience will be dramatically improved. For example, when a customer requests the opening of a new account or an overdraft facility, we won't hand him a blank form and ask him to fill it out; the system will generate a form with all of his details from his customer record and the customer will only need to sign the form. Within two to three years, even this step won't be necessary as the customer will simply view the product agreement on screen and authorise the new account with a biometric confirmation or on-screen signature through touch screen capability.

Perhaps the biggest single improvement will be in telephone-based service technologies. Already we are starting to see the introduction of **voice recognition systems** in IVR systems. As we move into video call technology, we will pair this with avatars to provide a more human

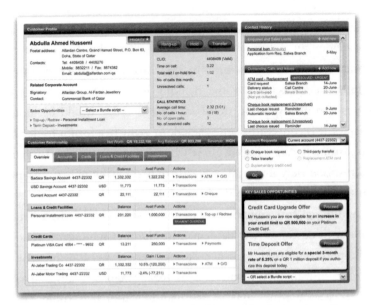

Figure 14.2 Contact centre dashboard for better in-call customer management (Credit: UserStrategy, Health Wallace)

interaction. However, even without video avatar agents, the voice recognition capability will provide a challenge for the bank to understand the most dominant enquiries through the phone channel and to optimise around these. In fact, channel optimisation will involve a much clearer understanding of which transactions take place across which channel, and improving the channel to cater for that process.

Rather than have an IVR system that sounds like we are reading out the bank's organisation chart (Press 1 for Retail Banking, Press 2 for Corporate Banking...), the IVR will be optimised for each customer, that is, he can either ask for exactly what he wants, or he will be presented with options most regularly chosen by him in previous interactions with the bank.

Channel and distribution

Channel management is probably the most significant change both operationally and technically. Currently, banks are sticklers for maintaining multiple technology and operational silos around customer channels. This situation is untenable as we move into the networked economies of the 21st century. Why? There are two significant reasons.

Firstly, customer channels will continue to evolve and emerge, presenting opportunities. For example, the app phone has generated a channel that is ubiquitous, but it is currently one that is not utilised by most banks. Skype and Instant Messaging are everywhere, with billions of users globally, and yet you can't contact your bank through these communication mediums.

Secondly, bank management will increasingly face the question of who should own all these new channels. In reality, a customer-focused channel team could deal with all these new opportunities agnostically without having to offset investment in one channel against some pre-existing budget thinking or bias within the marketing and IT teams, for example.

To illustrate, Mobile TV is already growing in popularity at an alarming rate across Asia, with Japan and Korea investing significantly in the technology that has already seen strong customer adoption. However, there is currently no mechanism within banks to generate content for this format—and just retasking TVCs is not an option. If Mobile TV was to be utilised as an engagement channel, where would the channel and content responsibility lie? Who would own it?

Again to illustrate, as point-of-sale technology evolves, currently this is likely to be compartmentalised as a payment solutions issue and relegated to MasterCard, Visa, Diners and Amex to solve, but your bank actually needs to find a way to optimise the point-of-sale experience for each individual customer. Who would own that? Most likely your bank would relegate this to credit card "usage" and leave it to the cards team. However, creating the right precognitive service selling offers requires more than a cards team. It requires a deeper understanding of customer behaviour through analytics.

All these challenges cannot simply be met by the current technology platform and organisation structures that most banks employ. How will the platform be optimised to serve a true multichannel services concept?

Figure 14.3 overleaf illustrates the complexity most banks find themselves saddled with today—multiple channels, largely married with current systems on a case-by-case basis, with independent technology bridges and interfaces. In addition, content that is created or published currently is purpose-built for one channel only each time it is required.

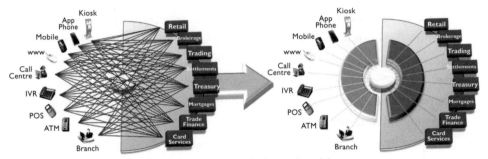

Figure 14.3 Bank 1.0 channel architecture
to BANK 2.0 channel architecture

In future, ATMs, phones, the Internet, mobile Internet and app phone devices, in-branch systems and the call centre will all leverage sales offers generated by the customer dynamics team. Just imagine publishing that content across these channels with eight or nine different content and channel platforms.

Consider the development effort required every time we find a channel is not adequately configured to handle the latest capability of the end user's device. Optimised channel management goes hand in hand with better content, service and experience. This is a goal that cannot be achieved in the current IT environment without considerable expense that is largely avoidable if an overarching team is created for customer sales offers. This streamlined environment will also enable the bank to respond to product opportunities in real time across every channel, rather than face production lags due to silo-ing.

Organisational Impact

As an organisation, your bank is rapidly going to have to re-engineer itself around the customer more effectively. The current departmental structure effectively creates competition for resources that could otherwise be optimised in servicing the customer. For example, removing duplication of silos such as the call centre, Internet banking and email handling for different business units (i.e. retail versus commercial banking, personal Internet banking versus commercial internet banking, etc). It means creating a consistent service across all channels because branch performance is no longer the key measure for customer experience. Customers now evaluate the bank on its performance across EVERY channel—a great branch experience will not save you if your Internet thingy sucks. Here are the specific likely changes.

Channel development to channel management

It does not make sense from a platform point of view to have separate channel infrastructure for commercial and retail banking services. For example, the skills required to manage the Internet channel in terms of things such as content management, search engine optimisation, engaging vendors and technical implementation are identical whether for commercial content or retail content. It makes no economic sense to keep the management of these areas separate. Likewise for call centre and email handling.

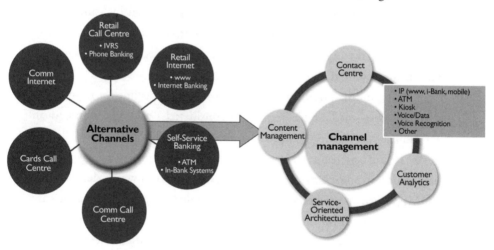

Figure 14.4 From channel silos to channel management

Increasingly your bank will need to find a mechanism for dealing consistently with enquiries and support/service requests across ALL channels. So if someone enters a branch, the teller or RM (relationship manager) will know that two days ago the client rang the call centre about the loss of his credit card. The RM can then see if the card is ready for pick-up without prompting from the client. If a service matter remains unresolved, when a caller rings the contact centre, the CSR (customer service representative) will be ready to resolve this call because the system has anticipated this and presented the options on the screen.

The added advantage for the customer will be that he will have access to all his related accounts aggregated through one contact centre and will rely on the bank to direct him to the appropriate solution, rather than having to call different numbers.

Consistency of service will generate brand collateral through a better overall service experience. That is, the Internet banking channel, cheque deposit machine, ATM, email, branch, and all the channels will be integrated so the message is consistent and no opportunity for service is lost due to technology or channel silos.

Marketing department transitions to customer dynamics

Most marketing organisations, including retail financial institutions, still have not optimised their marketing approach to the digital and interactive mediums. Marketing teams are heavily geared towards broadcast or, as Seth Godin characterised, "interruption" marketing.

While so-called traditional media have been in decline for most of the decade, "'new' media have been often considered a mystery or an add-on for campaigns targeting pimply students and "World of Warcraft" playing geeks. It just isn't really taken that seriously. Additionally, management is demanding stronger accountability and strong metrics which demonstrate ROMI (return on marketing investment), while marketing departments are seeing CPM (Cost Per Impression) response rates and sales figures plummet off traditional advertising mechanisms that are losing their effectiveness.

What still invariably happens today, due to a deficient skill set within the marketing department, is that rather than coming up with original and innovative approaches to utilising social networking and digital technologies, most marketing departments are stuck and simply find themselves trying to retrofit traditional campaigns onto new media, with very little success.

Let me illustrate. **Electronic direct mail** is virtually useless as a medium now because such broadcast, one-size-fits-all email campaigns have been overused, creating a broad brush classification of all such email marketing as spam—and so traditionalists are arguing that direct email doesn't work. Well, of course not!. All the traditional marketers who have been dumping direct mail down their email lists have simply ruined the medium for marketing all together.

Traditional print ads that were poorly retrofitted online as so-called **superstitials and interstitials banners** are now blocked by pop-up blockers because they were intrusive, poorly positioned, lacked any sort of targeting and just got in the way of a productive Web content experience.

Rich media flash website introductions, designed by traditional advertisers in an attempt to create a 15- or 30-second TV commercial online, are "skipped" 100 per cent of the time because they just get in the way of where we want to go. If advertisers had just understood in the first place that the Web was the Web and not TV, we might have been able to use flash a little more judiciously and effectively for ads. Then we've got classified ads and mini-billboards in the form of banners that have become ineffective because customers increasingly filter them out—a psychological effect known as "banner blindness".

All this while, most traditional marketers simply have not understood that the Web is actually not media at all (at least not in the sense of the TV, newspaper, magazines, radio or billboards)—it forms part of a dialogue or "journey" by the customer that starts with an integrated campaign, a URL, a click, or a search engine, and eventually ends with the sale. Web and mobile are about an actionable process. They are about triggering interest, and resulting in an action, usually a lead or a direct sale. No other channel in the marketer's arsenal is so self-contained. Every other medium requires some other sort of separate and distinct action through another totally separate channel. So, for example, if you see a really good magazine ad that you'd like to respond to, you either have to pick up the phone, get in the car and drive down to the branch, or visit the website.

But if you see a really compelling campaign online, you just click and apply. Done! What was traditionally about brand messages through old media is all about an **interactive, brand experience** in the digital space. The fact is that you can't really create an experience on traditional media. I can hear traditionalists argue that TV and radio ads can create an experience. Perhaps, but an emotional response is not the same as an engagement experience which can result in an *immediate sale or acquisition* which online or mobile offers. The immediacy of the Web is a punishing environment; you need to deal with it in a completely different manner than for advertising. Thus, the concept of user interaction, versus "perception" or recall.

In the marketing environment of **BANK** 2.0, television commercials will be a thing of the past, with technology such as TiVo or just plain old P2P downloading eliminating the effectiveness of the medium.

Today, we need to target customers with pin-point accuracy with offers only directly relevant to their needs. Any brand that continues to force irrelevant offers using the broadcast or shotgun approach to lead generation that is common in methods such as direct mail, will see the effectiveness of those methods reduce from the current levels of 0.4 per cent response rates to nothing, zero, nada.

The organisation will need to adapt to this pressure by creating a team that focuses on constant optimisation of customer propositions through segmentation analysis, behavioural analytics, just-in-time product manufacturing and permission-based marketing. This will be supported by neural networks through precognitive selling, which will anticipate the sale based on behavioural patterns and segmentation data, thus generating average response rates of 20–25 per cent rather than the 0.01 per cent (TVC) to 0.4 per cent (direct mail) of traditional push approaches common today. Customers will perceive this not as marketing, but rather as **servicing** because the messages will be individual, unique and integrated seamlessly into their banking experience.

When I am shopping, my bank will offer me a line of credit to use to purchase the bedroom setting I'm considering buying. When I'm booking travel online, my bank will automatically provide me with travel insurance coverage at an agreed set rate. When I get a salary increase, the bank will automatically offer me an upgraded platinum credit card with an extended credit limit. Insurance on my home, my car, my boat will be integrated into a central policy, automatically updated unless I nominate otherwise—

I will only be asked the first time. I will be offered bundled products that are constantly optimised.

The customer is the primary focus, and brand recall is only relevant to customers who can be serviced in this way. The marketing team will be a true revenue generation platform, not through advertising, but through channel, customer and offer management.

The biggest obstacle the marketing team will need to face is the concept that the customer dynamics team is far more than a new method of organising advertising activities. It is a team dedicated to optimising and showing customers the brand experience, and not just telling them what the bank can offer.

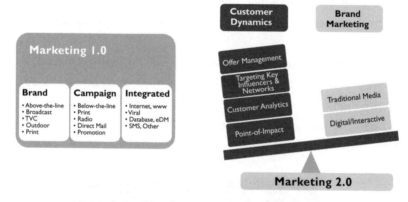

Figure 14.5 The marketing department will be completely deconstructed in BANK 2.0

Distribution and branch management

In the world of **BANK 2.0**, branches will be both more important and less important in the average customer relationship. Time-poor individuals in the High Net Worth and working professional bracket will increasingly look to relationships managed remotely, and once these are in place, will rarely visit the branch. Customers looking to do a major transaction, such as a mortgage contract, establish a new banking relationship or optimise their portfolio or credit footprint, will seek out assistance via specialist advisors. But low-involvement products and transactions will be totally relegated to more efficient channels.

In this environment, there will be four primary organisational units that replace what is now traditionally called branch management. Those areas will incorporate:

- **Branch (frontline) Management,** which manages the staff and resources of *physical locations* where customers will go to interact with a person.
- **Channel Management,** which will increasingly include *fully automated branches* such as ING Direct's Bank Cafés and other solutions incorporating avatar tellers and the like.
- **Customer Dynamics,** which will use the bank's *customer intelligence* capability constantly to optimise the customer experience through product offerings, sales campaigns and offer management, including the way staffed branches will utilise this data.
- A **Partner Management** team, which will increasingly extend your bank's capability to deliver content, product, information, solutions and presence to beyond the pure brand presence to point-of-sale and point-of-impact solutions.

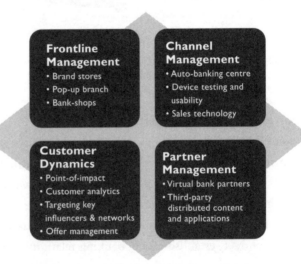

Frontline Management
- Brand stores
- Pop-up branch
- Bank-shops

Channel Management
- Auto-banking centre
- Device testing and usability
- Sales technology

Customer Dynamics
- Point-of-impact
- Customer analytics
- Targeting key influencers & networks
- Offer management

Partner Management
- Virtual bank partners
- Third-party distributed content and applications

Figure 14.6 Four key support teams for frontline management

Change management and transition

These changes could occur incrementally if we had time. But we don't have the time because customer behaviour has already irrevocably morphed. The key constraint on true innovation and growth for the 21st-century bank, however, will be mostly internal politics based on current departmental hierarchies and trying to get this organisation structure to adapt.

For example, who should the CEO appoint to control the new Customer Dynamics capability? Should it be marketing, the branch team, or an entirely new team with support from the business, marketing personnel and branch advisors?

Is Channel Management an IT function, a marketing function, or does it become a parallel to branch operations at the same organisational level? The latter is most likely as channel management will actually overtake branch operations in terms of revenue and metrics weighting within the next two years, if it hasn't already at your bank.

Automated channels could account for more than 70 per cent of revenue within the next five years most certainly, if not sooner. Having your primary revenue capability as a subset of transactional services, marketing or IT simply doesn't make strategic or business sense. In the most fundamental sense, Channel Management and Customer Dynamics will become the most strategic business units for the retail bank—and they don't even exist today.

Of course, if any one of the existing primary business units claim channel management as a subset of their existing departmental structure, the impact will simply be that organisational improvements will be stalled as non-aligned resources without the required skill sets are rapidly sequestered to a team that requires a completely new process and approach to customer fulfilment.

Compliance and Legal are also increasingly going to challenge our capability to service customers efficiently. In the **BANK** 2.0 and post-GFC (global financial crisis) economies of the world, customers are frequent users of technology where they can get results instantly. The following are examples of where we can get instant product fulfilment online: airline tickets, car rentals, hotel bookings, movie tickets, music and movie downloads (and purchases), serviced apartment rentals, office services, personal loans, insurance policies and insurance claims.

Increasingly, customers will expect Straight-Thru Processing and immediate approvals on everything from a new credit card, credit limit upgrades, line of credit or overdraft facilities to tax loans, personal loans and even a mortgage. In this environment, your bank will need to accept that we get only two chances to get the required information from customers—the first time they apply for a product with us, and then incrementally over time. We will never again be able to ask them for the same information twice, or the customer will punish us.

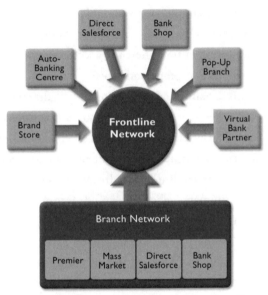

Figure 14.7 No longer one-size-fits-all branch banking

The bonus question

You would think that with all the government moves to restrict or tax big bonuses in the banking arena, and the massive public outcry by consumers over huge fees and bailout money seemingly funding the bonuses, bankers might take a pause before agreeing to big financial rewards this year. Unfortunately, for banks and customers alike, this seems unlikely.

In news that seems to defy all sense of decency, JP Morgan is expected to reward their executive team with more than $29 billion in bonuses in 2010 after a resurgence in investment banking during Q4 of last year. The payout will represent an increase of 28 per cent over last year's figures. To top it off, Jamie Dimon, JP Morgan's chairman and CEO, has been very vocal about his opposition to the 50 per cent windfall tax on bonuses, even threatening Chancellor Alistair Darling that if JP Morgan wasn't exempted, they might drop the plans for their £1.5 billion European headquarters in London.[1]

Bank of America, on the other hand, perhaps believes that it is taking the high road by rolling back bonuses for their top executives. But I think they've missed the point.

> "When you look at the overall pool or the individual payouts, they will not be [at] record [levels] ... They will be up from last year, but last year was significantly depressed."
>
> BofA spokesperson Bob Stickler, January 2010

The banks, however, are not really on their own in the blame game. The first issue here is that the bailouts provided late in 2008 and some again in early 2009 taught the banks that they were essentially immune to the negative effects of their own strategy in relation to subprime and the securitisation of bad debt. Secondly, when they did receive massive bailout assistance, many of the banks didn't require the money but took it to reduce liquidity pressure and simply because they had access to cheap government money. A JP Morgan executive confided in me that even when JP Morgan first tried to pay back the TARP (Troubled Asset Relief Program) funds in early 2009 that they had availed themselves of, the US government wasn't interested and told them to hold on to the funds. *The Wall Street Journal* backed up this insider view when it reported that "...in June, several firms eager to escape government scrutiny were *allowed* to return their bailout cash to the government."[2]

The message appeared to be that the banks could do no wrong. They simply must have started to "drink their own kool-aid".

In all of this, I feel the banks have forgotten that their sole responsibility as an organisation is first to their shareholders, and secondly, customers. I'm sure shareholders at the moment would prefer to see the banks being a little conservative on the bonus front and either hanging on to cash or even distributing more dividends. But the biggest losers here are the customers.

At a time when bank fees are at record levels and banks are ready to jump on delinquent mortgage, credit card or loan issues lightning fast, their unwillingness to show even a modicum of restraint in respect of bonuses shows their overconfidence and lack of responsibility.

Banks could find themselves in significant trouble with shareholders over this issue. As bankers in the UK reacted with fury at the 50 per cent windfall gain tax imposed by Alistair Darling's team, they are yet again preparing at least £40 billion in bonuses during the first quarter, with the tax to be borne by the bank directly, thus reducing net profit and dividends. In a clever strategic move, the Chancellor has called on shareholders to take direct action in ensuring that bank executives are held responsible for their fiduciary duty to shareholders over the bonus tax.[3] There is also the story of an Illinois shareholder of Goldman Sachs who is taking the investment banking giant to court over the bonus scandals.[4]

Customers aren't impressed. As customers, we once again want to believe that banks are capable of acting in our best interests, and that as customers we are no longer simply a number that is just part of a bonus/profit making machine. With banks reining in expenses and cutting costs, it looks like some units of the majors will actually do pretty well in 2009. Thus, shouldn't they feel justified in taking a bonus? Well, given the groundswell of sentiment against the big banks, it would make sense for these executives to give some relief to their hard off customers first, before thinking of themselves and their own pay packets.

It looks like we'll have to wait a while longer for banks to have this epiphany.

BANK 2.0 Projects Roadmap

Programmes

The following table represents the probable timeline of changes required and the impact or base requirements of each programme initiative. There needs to be significant effort in creating a central innovation programmes office that will project manage these initiatives so that they don't get hamstrung by traditional departments arguing over turf and responsibilities.

Table 14.1 Probable timeline of changes required and their impact

Programme Initiatives	Structural Impact
Year 1	
Customer Analytics and Customer Data-Mart	Analytics will drive product development and marketing, and we will be channel agnostic. Every channel is equally important, branches are just one.
Contact and Content Management	Call centre, Web, email, branch enquiries are all supported through a central integrated system, content also similar. Contact centres will start to integrate IM and Skype-type technologies.
Real-time Dashboards	Product managers, channel managers and marketers will see in real time the impact on revenue that campaigns and products are having.
Mobile App Phone Support	Introduction of mobile app phone support for high traffic Internet banking and ATM equivalent functions (account balance, transfers, etc).
Social Media Support	Get a presence on every social media platform possible, and have customer advocates listening to everything customers say about your brand online and in forums.
Year 2	
Marketing Reform—Offers and Segmentation	Offers to customers will be individualised, designed as a journey moving from one channel to the next seamlessly; more than 50% of revenue will be from the Web and Web-enabled phones. Marketing will be split into Brand Marketing (the current team), and a much larger Customer Dynamics team.
Straight-Thru Processing (STP)	General insurance, line of credit, loan and credit card applications will happen instantly without human intervention.
Credit Risk Management	In line with STP, we need to automate credit risk assessment.
Strategy	Budgets will be dramatically re-engineered to support the Internet and mobile as the primary channels, and interactive marketing as the primary revenue triggers. Branch expenditure will be level with Internet and mobile spend.

(cont'd on the next page)

Table 14.1 (cont'd) **Probable timeline of changes required and their impact**

Programme Initiatives	Structural Impact
Year 3	
Payments Technology	Mobile payments, e-wallets, etc. will mean much more flexible point-of-sale capability outside of the branch.
Partner Strategy and Distributed Applications	We'll start with the first elements of partner branches and engaging customers where the point-of-impact is, rather than relying on them coming to us.
Marketing Reform— Customer Dynamics and Permission Marketing	TVC, print and direct mail will make up only 25% of the total marketing budget. Rather than the current broadcast/ push approach, we will ask and capture customer needs at every opportunity so that we only offer relevant and timely product.
Year 4	
Branch Automation	We'll see more automation in the branch as we steer customers away from transactional behaviour at the counter, to a sales and service centre. **Cheques and physical credit cards are being phased out.**
Voice Recognition and Avatars	"Press 1 for English" will be replaced by spoken commands and the first video avatar support personnel will start to appear.
Year 5	
Distributed Banking	Branches will have either started to consolidate into megastores (where we educate, inform and sell), or targeted service banking will take place anywhere customers are, with mini-branches and bank-shops the norm.
Non-Financial Metrics	The new measures of financial success will be customer profitability, Lead to Offer Ratios and Return on Marketing Investment.
Predictive Marketing	New POS and cloud computing technologies will enable location-based and point-of-impact marketing offers in real time to customers wherever they are, but the offers will need to be constructed based on predictive models driven for detailed customer analytics.

Projects

Here are projects your team can start today to improve your readiness for the full impact of **BANK** 2.0. Each initiative is designed to create fast payback revenue and service opportunities.

Table 14.2 **Projects to get banks ready for full impact of Bank 2.0**

Project/Initiative	Desired Outcome
Mobile App Development	This is not one application (i.e. mobile banking), but looking at continual opportunity for a set of mobile banking applications that your customers can utilise. The primary m-bank apps (account balance, transfers, pay bills) need to be launched by mid-year 2010 latest.
Widget Development	Widget development for social networks and for desktops needs to be attacked with the same haste as mobile. In fact, it is possible that you can share code or duplicate the apps for both platforms.
Twitter and Social Networking listening	Put members of your customer service team on this today. Search out social network groups, tweet feeds and others to see what customers are saying about your bank. Then go fix the problem.
In-Branch Cash/Cheque Deposit Machines	Reduce over-the-counter transactions that are purely cost for the branch.
Branch Meeters/Greeters	Redirect non-optimal transactions to self-service automated capability.
Customer Analytics	Improved behavioural analytics on customers across all channels to understand better which "tasks" customers prefer to do in-branch, versus online, etc.
Sales Intelligence and Automated Offer Capability	Real time and precognitive offer management for existing customers delivered in the form of prompts, offers or service messages.
Branch Customer Dashboard	Customer information dashboard that shows entire relationship footprint at a glance, along with current risk rating, credit approvals and suggested sales offers.
Improved Staff Mobilisation	Focused service and sales training programmes, along with better KPIs that focus on more than simply number of applications per month, or total revenue.

(cont'd on the next page)

Table 14.2 (cont'd) Projects to get banks ready for full impact of Bank 2.0

PROJECT/INITIATIVE	DESIRED OUTCOME
Business Process Re-engineering on select processes	Reduction of layering between sales and service departments, including the removal of duplicate "skills" within "competing" product units. Creation of "customer dynamics" capability as owners of customers, rather than product competing for revenue from same.
Straight-Thru Processing and Credit Risk Management systems	Enabling customers to get immediate fulfilment for an application rather than waiting the obligatory 24, 48 or 72 hours later due to antiquated manual or human "processes" in the back office. Results in improved service perception and reduction of abandonment due to ongoing process demands (i.e. proof of income, faxing of three months' bank statements, salary certificate, etc.). Additional benefits include reduction of compliance errors through manual mishandling.
Customer Friendly Language Initiative	Use of ethnography, usability research, audits, customer focused observational field studies and focus groups to improve language and simplicity of application forms and communications with customers within branch (and beyond).
Contact Centre Apprenticeships	Start new bankers off in the call centre for three months. Help them get a feel for the issues customers face from the get go. Put line managers in the call centre for one week every year.
Homesourcing Trial	Consider trying homesourcing as a way of better contact centre staff retention.
IM/VoIP Integration to Contact Centre	Find a way to integrate Instant Messaging and Skype into your contact centre.
IVR System Redesign	Think about prioritising menu options based on traditional traffic analytics, thus reducing IVR navigation for those calls that are most frequently made. Think about the use of emotive voice recognition technology to redirect upset customers to a "customer advocate".
Upgrade Your Service Culture	Create a service culture within the bank that gives contact centre staff pride in their role within the customer equation.
Compliance and Legal Metrics Update	Give the compliance, risk and legal departments metrics which measure how many process problems they resolve through proactive consultation, rather than how many road blocks they assemble.
Usability Tests of all websites	Assess any issues with current website language, layout, design and process.

Table 14.2 (cont'd) Projects to get banks ready for full impact of Bank 2.0

Project/Initiative	Desired Outcome
Content Management Systems	The old dot.com favourite is back, but this time enabled across the organisation so we can "publish" new content continuously. The best analogy is to imagine that your bank is publishing a product catalogue and investor information magazine based on your product daily to customers. Also commit to XML and CSS as core technologies on the interface, along with considering more IP based backbone.
Search Engine Optimisation of websites	Organic search engine optimisation should be the strategy of every institution, but it requires rethinking what content you actually put on the site because it needs to be driven by what customers are actually looking for.
ATM Advertising revamp	Consider wraps, on-screen video, coupons and other advertising methods for some revenue opportunities on this old chestnut.
Better ATMs	Look at gradual replacement of ATMs to incorporate touch screen, Windows platform, and better usability and efficiency. Incorporation of personalisation functions and reduced screen flow should be key goals. Recurring transactions and bill payments can be saved for later recall to further improve speed.
Biometrics on ATM	When will be the time to get rid of PINs so we can have a truly secure ATM platform?
Mobile Payments NFC proof-of-concept	Start testing prototypes of debit card phone proxies. It won't be long before your customers' phones replace the bit of plastic in their wallet. Be prepared for this when it happens.
RFID recognition	Using RFID ATM cards or RFID-enabled app phones, build branch capability to recognise a customer when they walk in the door, and have the branch dashboard ready with their profile and a possible offer before they even approach the low-counter specialist.
Mobile Remittance Integration	Work with the foreign currency exchange providers and with employers of overseas foreign workers to give the unbanked access to mobile remittance payments P2P.
P2P Payments capability	Look at P2P payments via mobile devices as a growth opportunity and a way to reach new customers.

The chart below represents the timeline of disruptive changes, along with the required organisational responses.

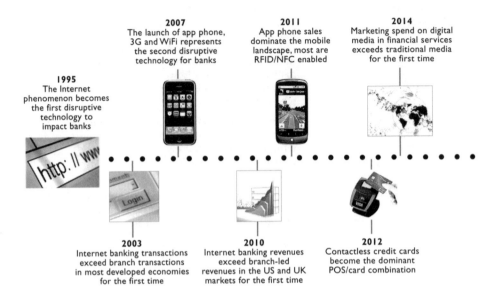

1995
The Internet phenomenon becomes the first disruptive technology to impact banks

2007
The launch of app phone, 3G and WiFi represents the second disruptive technology for banks

2011
App phone sales dominate the mobile landscape, most are RFID/NFC enabled

2014
Marketing spend on digital media in financial services exceeds traditional media for the first time

2003
Internet banking transactions exceed branch transactions in most developed economies for the first time

2010
Internet banking revenues exceed branch-led revenues in the US and UK markets for the first time

2012
Contactless credit cards become the dominant POS/card combination

Conclusion

BANK 2.0 is about change. Change that is inevitable, change that is speeding up, and change that is extremely disruptive. You may have read this book, and you might not agree with all the predictions. You may not think that cheques, credit cards and cash are under threat from new technologies. You may not feel that the marketing team needs to change its approach significantly because it has worked so well in the past. You may feel that the bank is in a strong financial position, so this is simply "much ado about nothing".

However, you may also agree that customers seem to be changing the way they engage banks at an annoyingly rapid rate. You are probably amazed at how many people are carrying CrackBerrys, iPhones, NexusOnes or Droids. You might be amused at how many people are discussing Facebook and Twitter, and you might be thinking that the global financial crisis has left your bank with a consumer confidence problem that is going to be extremely difficult to crack.

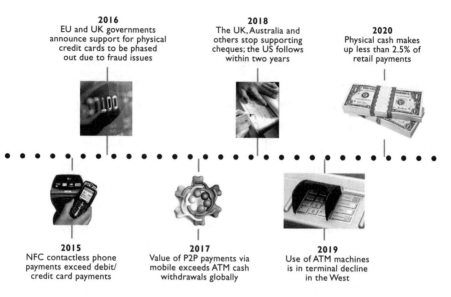

2016
EU and UK governments announce support for physical credit cards to be phased out due to fraud issues

2018
The UK, Australia and others stop supporting cheques; the US follows within two years

2020
Physical cash makes up less than 2.5% of retail payments

2015
NFC contactless phone payments exceed debit/credit card payments

2017
Value of P2P payments via mobile exceeds ATM cash withdrawals globally

2019
Use of ATM machines is in terminal decline in the West

In any case, I hope that the evidence and thinking presented in this book are enough to produce one thing—a groundswell of support within your bank to innovate and try new ways of reaching, engaging and communicating with your customers.

If you are a banker and you do nothing else as a result of reading this book, just keep asking customers and listening to customers about HOW they want to engage with the bank. When those customers talk about mobile, the Internet, social networks and so forth, don't say anything about the effectiveness of your branch, or how broad your ATM network is. Take it on-board and figure out how you are going to provide customers with accessibility and functionality on the right channels at the right time.

Look at the structure of your executive team and your marketing team today. If your executives aren't on Twitter, LinkedIn or Facebook, help them get engaged with these new tools to understand what your customers are doing. If your marketing budget is not at least 50 per cent dedicated to digital or interactive today, then replace your heads of marketing immediately because they are already five years behind the trend on this. Your organisation needs a dedicated and relenting focus on trying new ways to engage customers across new mediums. Most banks are already behind the eight ball on this.

But most critical of all—**innovate and experiment.** Things move so fast technologically these days that you cannot wait until a trend is three years into its cycle to adapt. Why not? Because by three years into the adoption cycle, the next big thing will already be on its way. The more channels, applications, touchpoints and locations you use to engage customers, the more you will discover that channel silos are crippling the speed with which you engage and convert sales opportunities. Start removing the silos immediately and revenue will go up. Create a team that is both an advocate for customers, and a team dedicated to creating the right offer, across the right channels, at the right time. Give these resources huge support because they are your new frontline. The customer dynamics team will be one of your largest departments within three years.

Technology is a means to an end, but it is also increasingly becoming a means to profitability. Channels are increasing in complexity, not decreasing. You need to manage the customer, agnostic of any one channel he or she chooses. The branch is no more important than any other channel; it is also not going to be the most profitable channel moving forward. So any business case you present is going to have to figure out how to tackle this channel/resource conflict sooner or later.

Understand one thing. Customers are not going back to the old ways of banking. They are moving forward. If you are not moving forward with them, then they will pass right by you—at warp speed. **BANK** 2.0 **is now.** Get moving, or get out of the way!

Endnotes

1   "Record bonus pot at JP Morgan", *Daily Telegraph*, 9 January 2010

2   "BofA Seeks to Repay a Portion of Bailout", *The Wall Street Journal*, 1 September 2009

3   "Treasury tells shareholders to block bank bonuses", *The Guardian* UK, 9 January 2010

4   "Goldman Sachs in the Firing Line", *Pittsburgh Tribune,* 10 January 2010

Part04

Glossary of Terms, References and Works Cited

Glossary of Terms

Adoption Rate how quickly it takes new technologies to be adopted by the public at large

ATM Automated Teller Machine; typically automated cash dispensing and depositing machine that emerged in the late 70s and early 80s

AML Anti-Money Laundering; the efforts through legislation, regulation and through systems to track, identify and stop the laundering of illicit funds into the mainstream banking system

App Phone a phone that provides open application support not limited to the phone handset, manufacturer's operating system and applications; most common instances are the iPhone, Droid and NexusOne

AR Augmented Reality; the term for real time digitally enhanced interactions with the physical real world environment

AuM Assets under Management; the term used to signify the sum total of assets held for a customer under management by a financial institution

Avatar a computer user's representation of himself/herself or alter ego for use on computer systems

B2B Business-to-Business; as in intraorganisational communication, collaboration and commerce; normally electronic, and usually using websites and/or Web services

Basel II the second of the Basel Accords, which are recommendations on banking laws and regulations issued by the Basel Committee on Banking Supervision

Blog a contraction of the term "Web log"; a type of website, usually maintained by an individual with regular entries of commentary, descriptions of events, or other material such as graphics or video

BPO Business Process Outsourcing; the practice of outsourcing some or all of the business's back-office processes to an external company or service provider; common with call centres and IT support

BPR Business Process Re-engineering; re-engineering business processes to either reduce costs or improve the flow of a process for customers

CLID Caller Line Identification; a system that identifies a customer based on the phone number they use to call a service provider

Cloud computing an emerging computing technology that uses the Internet and central remote servers to maintain data and applications

CME, HKEX, LSE, KOSDAQ, NASDAQ abbreviations for various stock exchanges around the globe

CPM Cost Per Impression; in online advertising, it relates to cost per (thousand) impressions

CrackBerry a fun name for the 'BlackBerry' series (RIMM/BlackBerry) of mobile phones

Cross-Selling a method of targeting and selling new products to an existing customer

CRM an abbreviation used to identify the system or management undertaking for Customer Relationship Management; sometimes also used as the abbreviation for Credit Risk Management

CSR Customer Service Representative; staff who work within the call centre to assist customers with enquiries

CTI Computer-Telephony Integration or Interface; a system that integrates telephone systems with computer networks

ECN Electronic Communications Network; an electronic network that facilitates trading between stock or commodities exchanges

Facebook a hugely popular online social network founded in 2004 for helping friends stay in touch and share information

FAQ Frequently Asked Questions; questions asked frequently by customers and put on the company's website to expedite answers

FMCG Fast Moving Consumer Goods; products that are sold quickly at relatively low costs

GPRS General Packet Radio Switching; a packet oriented mobile data service available to users of the 2G and 3G cellular communication systems in global system for mobile communications (GSM)

GSM Global Systems for Mobile Communications; the primary standard for digital mobile phones in use by 80 per cent of the global mobile market

Haptic Touch technology that interfaces with the user through the sense of touch

High-Counter the typical teller station within a branch for conducting over-the-counter transactions

HNWI High Net Worth Individual; the most attractive client segment for retail banks; HNWIs typically invest US$150,000–US$1 million in investment type products

IM Instant Messaging; a protocol for communicating between two parties using text-based chat through IP-based clients

IxD Interaction Design; a customer-led design methodology for improving the interaction between customers and a systems

IP Internet Protocol; the primary protocol for transmitting data or information over the Internet

ISP Internet Service Provider; a company that provides Internet access to customers

IVR Interactive Voice Response (systems); the automated telephone support systems you hear when you call a 1-800 help line or customer support number that uses menus and responses via touch-tone and/or voice response to navigate

KPI Key Performance Indicators; metrics (or measures) used within the corporation to measure the performance of one department against another in respect to things such as revenue, sales lead conversion, costs, customer support, etc

KYC Know Your Customer; an internal compliance regulation to ensure the accurate identification and validation of a customer and understanding his transactional behaviour

LAN Local Area Network; a computer network covering a small physical area, such as a home, office, or small group of buildings

LED Light Emitting Diode

Low-Counter typically a desk station within a branch where the relationship manager can sit with customers and potential clients and advise them on available products and services

Lo-Fi Prototype a simple method of prototyping products, interfaces or applications and testing with target customers or users

LinkedIn an online social network for business professionals

MFI MicroFinance Institution; an alternate form of bank found in developing countries which provides microcredit lending

Mobile Portal a website designed specifically for mobile phone interfaces and mini-browsers

M-PESA, GCASH, T-Money, Edy, Suica mobile phone-based technology for payments between two parties

NFC Near Field Communication; a short-range high frequency wireless communication technology which enables the exchange of data between devices over about a 10-centimetre distance

OLED an organic light emitting diode; also organic electro luminescent device (OELD), a LED whose electroluminescent layer is composed of a film of organic compounds

OTC Over the Counter; refers to physical transactions or trades done on behalf of a customer by a trader or customer representative who has access to a specific closed financial system or network

PPC Pay-per-Click; a method of paying for appearing in search engine results by bidding and paying for specific keywords; you then pay at the successful bid rate every time a user/visitor clicks on your link

P2P Peer-to-Peer or Person-to-Person; a method of passing information or data via IP-based communication methods between two individuals connected to the Internet via computer or mobile devices

POS Point-of-Sale; the location where a retail transaction occurs; a POS terminal refers more generally to the hardware and software used at checkout stations

Prosumer a portmanteau formed by contracting either the word professional or producer with the word consumer; in respect of this publication, it identifies the role of the modern consumer of content who is also a producer of content on, for example, YouTube, Facebook and Twitter

PSTN Public Switched Telephone Network; the traditional copper-wire and exchange based landline telephone system

RFID Radio Frequency Identification; a short-range radio communication methodology that uses "tags" or small integrated circuits connected to an antenna that when passed within the range of a magnetic reader, is able to send a signal

RM Relationship Manager; a dedicated customer service manager assigned to look after specific customers, usually High Net Worth ones

ROMI Return on Marketing Investment

SDK Software Development Kit; a package provided by a mainstream software or operating system provider to the developer community to assist them with application construction

SEO Search Engine Optimisation; the science of optimising websites so that they appear in the top results for search engine enquiries

SIM Card subscriber identity module (SIM) on a removable SIM card securely stores the service-subscriber key (IMSI) used to identify an individual subscriber on a mobile phone

SMS Short Message Service; a system of communicating by short messages over the mobile telephone network

Snail Mail the term used by proponents of digital technologies to describe traditional mail and the postal system

Spam unsolicited bulk email sent out simultaneously to thousands or even hundreds of thousands of email addresses to promote products or services

Stored-Value Card monetary value stored on a card not in an externally recorded account; examples are the Octopus, Oyster and Suica systems used to replace public transport ticketing

STP Straight-Thru Processing; the implementation of a system that requires no human intervention for the approval or processing of a customer application or transaction

TiVo a brand and model of digital video recorder (DVR) available in the US, UK, New Zealand, Canada, Mexico, Australia and Taiwan

Touchpoint any channel or mechanism by which a consumer has day-to-day interaction with a retail service company, such as a bank, in order to transact or conduct business

TVC the industry abbreviation for television commercials

Twitter a social media website that supports microblogging between participants in the network; sort of like an SMS broadcast system for the Web

Up-Selling a system of selling an additional service at a higher margin or total revenue within the same product or asset class to a customer, typically upgrading from one class of product to another

URL Uniform Resource Locator; an "address" or identifier that is used to locate and retrieve documents hosted on the World Wide Web

UT Usability Testing; the science of testing how users interact with a system, product or interface through observation

VBC Video Banking Centre (Citibank, circa 1996); an interactive, 24-hour personal banking centre providing access to personal banking experts, 24 hours a day, seven days a week, through integrated voice, video and data connection

Virtual Currency currencies such as Linden dollars, QQ coins, Project Entropia Dollars (PED), etc. that exist in the virtual world and can be exchanged for real currency by users

VoIP Voice over Internet Protocol; an Internet-based protocol that allows users to use voice communication such as over a telephone system

VSC Virtual Support Centre; a call centre virtually supported by customer service representatives who typically operate from home (i.e. homesourcing)

WAP Wireless Access Protocol; the original protocol for simple Internet browsing or simple menu interactions via 2G (Digital) mobile phones

Web 2.0 the term commonly associated with Web applications that facilitate interactive information sharing, interoperability, user-centred design and collaboration on the World Wide Web

Widget a generic type of software application that is usually portable and works across different operating systems and devices

WiMax Worldwide Interoperability for Microwave Access; a telecommunications technology that enables wireless transmission of data from point-to-multipoint links to portable and fully mobile Internet access

WTO World Trade Organization; an international organisation designed by its founders to supervise and liberalise international trade

XML eXtensible Markup Language; a set of rules for encoding documents electronically

References and Works Cited

Australian Broadcasting Commission. "Community banks silence sceptics". Retrieved from The 7:30 Report (18 August 2002). http://www.abc.net.au/7.30/content/2002/s679904.htm

Berghman, L., & P. Matthyssens, P. and K. Vandenbempt. "Building competencies for new customer value creation: An exploratory study". Industrial Marketing Management, Vol. 35, No. 8 (2006), pp. 961-73.

Cavell, D. J. *The Branch is Bank: Global Case Studies in 21st Century Banking Success*. London, United Kingdom: VRL Financial News (2008).

Colurcio, M., & C. Mele. *La generazione delle idee. La gestione dei percorsi di innovazione*, Giappichelli, Torino (2008).

Dan Milmo, J. T. "Treasury tells shareholders to block bank bonuses". *The Guardian*, 9 January 2010.

Federal Reserve Bank of Philadelphia. *An Examination of Mobile Banking and Mobile Payments: Building Adoption as Experience Goods.* Philadelphia: Federal Reserve Bank (2008).

FinExtra. "Philippines mobile phone-based microfinance bank set for launch" (13 October 2009). From Finextra.com: http://www.finextra. com/fullstory.asp?id=20598

First Data/Tower Group. *The Risks and Opportunities in a Mobile Commerce Economy.* First Data/Tower Group (2008).

Fitzpatrick, D. "BofA Seeks to Repay a Portion of Bailout". (Online, Ed.) *The Wall Street Journal*, 1 September 2009.

Gilb, T. "Usability is good business". In T. Gilb, *Principles of Software Engineering Management* (20th ed., Vol. 1, p. 464). Ormerudsveien, Kolbotn, Norway: Addison-Wesley Longman (1988).

Internet Advertising Bureau. *IAB Internet Advertising Revenue Report.* New York, NY, USA: PricewaterhouseCoopers, New Media Group (2009).

Lomas, N. "Barclaycard and Orange unwrap contactless credit card" (7 January 2010). From Silicon.com: http://www.silicon.com/technology/ mobile/2010/01/07/barclaycard-and-orange-unwrap-contactless-credit-card-39744115/

Lunden, I. "Murdoch Paper Blocks UK Aggregator Before Paywall Goes Up" (8 January 2010). From Paidcontent.org: http://www.creditcards. com/credit-card-news/credit-cards-of-the-distant-future-1273.php

Markowitz, J. "Illinois little guy takes on big stink at Goldman". *Pittsburgh Tribune-Review*, 10 January 2010.

Mbugua, J. "Big Banks in Plot to Kill M-Pesa". *Nairobi Star*, 23 December 2008.

McDonald, J. "What will credit cards look like in 25, 50 or 100 years?" (17 February, 2009) From CreditCards.com: http://www.creditcards. com/credit-card-news/credit-cards-of-the-distant-future-1273.php

Melouney, Carmel . "BlackBerry users becoming addicted to gadget". Sunday Telegraph, 11 May 2008. Retrieved from http://www.news.com. au/technology/story/0,25642,23676081-5014108,00.html

Parliament of New South Wales. Bank Branch Closures. Full Day Hansard Transcript (15 October, 1998). Sydney, NSW, Australia: Parliament of New South Wales.

Pisani, Joseph. C. "Workplace BlackBerry Use May Spur Lawsuits" (9 July 2008). Retrieved from CNBC: http://www.cnbc.com/id/25586129

Quinn, J. "Record bonus pot at JP Morgan". *Telegraph (UK)*, 9 January 2010.

Reserve Bank of Australia. "Bank Fees in Australia". Melbourne, Australia: Reserve Bank of Australia (1999–2009).

Sang-Hun, C. "In South Korea, All of Life is Mobile". *International Herald Tribune* (Technology), 25 May, 2009, p. 1.

Sophos Security. "50% of employees blocked from accessing Facebook at work, Sophos survey reveals" (21 August 2007). From Sophos.com: http://www.sophos.com/pressoffice/news/articles/2007/08/block-facebook.html

Szuc, G. G. The Usability Kit. 397. Melbourne, Vic, Australia: Sitepoint (November 2006).

UK Payments Organization. "Statistical release – 30 November 2009". London, United Kingdom: APACS (2009).

User Strategy Ltd. *Global Internet Strategy and Online Survey*. Standard Chartered. Hong Kong: User Strategy, (2007)

User Strategy Ltd. (2003). *Wealth Management Usability and Interaction Study*. HSBC, Electronic Channel Development. Hong Kong: User Strategy Ltd., (2007).

UserCentric.com. "Early Adopter iPhone User Study Identifies Baseline Issues with iPhone Interface" (12 July, 2007). Retrieved March 2009 from UserCentric.com: http://www.usercentric.com/news/2007/07/12/ early-adopter-iphone-user-study-identifies-baseline-issues-iphone-interface

World Bank Organization. *Migration and Development Brief (11).* Washington, DC, USA: World Bank (2009).

Wurster, P. E. *Blown to Bits: How the New Economics of Information Transforms Strategy.* Harvard Business School Press, (2000).

References and Case Studies concerning HSBC Asia Pacific and HSBC Hong Kong provided with the permission of HSBC Banking Group (HK) Ltd.

All references to Reserve Bank Australia Bulletins and data with the permission of Reserve Bank Australia.

iPhone™ and the App Store™ are trademarks of Apple, Inc. iPod touch® and iTunes® are trademarks of Apple, Inc., registered in the US and other countries.

Other References

electronic trading. In *Encyclopaedia Britannica.* Retrieved 16 July 2008, from Encyclopaedia Britannica Online: http://www.britannica.com/ EBchecked/topic/183888/electronic-trading

Wikipedia

About the Author

BRETT KING is strategic advisor to the financial services sector and founder of the International Academy of Financial Management, a  professional association focused on financial services. He is an International Judge for the Asian Banker Retail Excellence Awards and for the Middle-East Business Achievement Awards.

A regular speaker at the top global conferences for financial services, Brett is an ackowledged expert on wealth management, customer experience and retail channel distribution strategy. He publishes regularly in his role as industry advisor on *Huffington Post* (Business News), Internet Evolution and his own personal blog, Banking4Tomorrow.com.

Brett also runs UserStrategy, a boutique consultancy focused on improving customer interaction for leading financial services companies and businesses. His clients include HSBC, Citigroup, UBS, Standard Chartered, Abu Dhabi Commercial Bank, Emirates NBD, BNP Paribas and many more. He previously led the Asia division for Modern Media/Digitas (part of the Publicis group) and the E-Business service line for Deloitte (Financial Services Industry focus).

Contact Brett King at:
E: bking@userstrategy.com
W: userstrategy.com

BANK 2.0